D1247035

PHYSIOLOGICAL CHEMISTRY

A Series Prepared under the General Editorship of

Edward J. Masoro, Ph.D.

I

PHYSIOLOGICAL CHEMISTRY OF LIPIDS IN MAMMALS

II

PHYSIOLOGICAL CHEMISTRY OF PROTEINS AND NUCLEIC ACIDS IN MAMMALS

In preparation

PHYSIOLOGICAL CHEMISTRY OF CARBOHYDRATES IN MAMMALS

ENERGY TRANSFORMATIONS IN MAMMALS

ACID-BASE HOMEOSTASIS:

Its Physiology and Pathophysiology

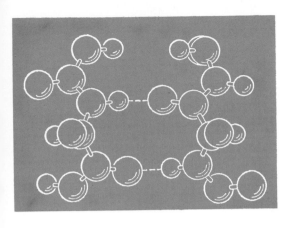

PHYSIOLOGICAL CHEMISTRY OF PROTEINS AND NUCLEIC ACIDS IN MAMMALS

George Kaldor, M.D.

Professor of Physiology and Biophysics,
Woman's Medical College of Pennsylvania

1969

W. B. Saunders Company

Philadelphia · London · Toronto

W. B. Saunders Company: West Washington Square
Philadelphia, Pa. 19105

12 Dyott Street
London W.C.1

1835 Yonge Street
Toronto 7, Ontario

Physiological Chemistry of Proteins and Nucleic Acids in Mammals

TO GLADYS
Julius and Nickolas

EDITOR'S FOREWORD

The past three decades or so have seen biochemistry emerge as possibly the most vigorous of the biological sciences. This, in turn, has led to a level of autonomy that has cut the cord linking biochemistry with its historically most important parent, mammalian physiology. For researchers in the fields of both biochemistry and physiology, this vitality has been most useful. But because of the arbitrary separation of these two disciplines in most teaching programs and all textbooks, the vast majority of students do not see the intimate relationships between them. Consequently, the medical student and the beginning graduate student as well as the recently trained physician find it difficult, if not impossible, to utilize the principles of biochemistry as they apply to the physiological and pathological events they observe in man and other mammals.

Therefore, this series is designed not only to introduce the student to the fundamentals of biochemistry but also to show the student how these biochemical principles apply to various areas of mammalian physiology and pathology. It will consist of five monographs: (1) Physiological Chemistry of Lipids in Mammals; (2) Physiological Chemistry of Proteins and Nucleic Acids in Mammals; (3) Physiological Chemistry of Carbohydrates in Mammals; (4) Energy Transformations in Mammals; and (5) Acid-Base Homeostasis: Its Physiology and Pathophysiology.

The series can be profitably used by undergraduate medical students. Recent medical graduates and physicians involved in areas of medicine related to metabolism should find that the series enables them to understand the theoretical basis for many of the problems they face in their daily work. Finally, the series should provide students in all areas of mammalian biology with a source of information on the biochemistry of the mammals that is not otherwise currently available in textbook form.

EDWARD J. MASORO

PREFACE

The purpose of this introductory text is to acquaint the reader with some of the important physiological functions of proteins and nucleic acids. The guiding principle in the organization of the first eight chapters is the assumption that the unique macromolecular structure of proteins and nucleic acids is the most important aspect of their physiological properties. For this reason the chemical basis of their macromolecular structure is presented first. This is followed by a brief description of some of the important physiological functions of these macromolecules, e.g., antigen and antibody structure and interactions, coagulation of blood, and the mechanochemistry of muscular contraction. In the interest of brevity only the basic principles of the intermediary metabolism of proteins and nucleic acids are considered in Chapters 9 and 10; it is my belief that this coverage provides the biologist and physician with background adequate for his usual needs and also with a basis for reading more deeply in this area should the need arise.

I am greatly indebted to Dr. Edward J. Masoro for his most valuable suggestions and discussions and most of all for his thoughtful editorial work on the manuscript. I am also very grateful to Miss Maureen Mullen and Mrs. Fran Wilderman for preparing the manuscript and to the editorial staff of the Saunders Company for their skill and endurance in editing this text.

GEORGE KALDOR

CONTENTS

CHAPTER 8

CHAPTER 9

CHAPTER 10

1

PROTEINS: A GENERAL DISCUSSION

Proteins are high-molecular-weight substances essential for both the structure and function of every living cell. Their biological roles are manifold, and many of them will be discussed in this text. A list of important examples of the biological functions of proteins would certainly include their role as catalysts (i.e., enzymes), as hormones, and as structural components of cells, as well as their function in ion transport, acid-base homeostasis, muscular contraction, and antibody-antigen reactions.

It is astonishing that a single class of compounds is capable of participating in such diverse biological functions. Of course, a very large number of different proteins are present in any organism. The intriguing fact is, however, that the fundamental structure of each of these vast numbers of proteins involves the combination of 20 amino acids in a unique manner; i.e., amino acids are the building blocks of all proteins. Proteins are heteropolymers of amino acids. The amino acids are linked together by the peptide bond, in which the carboxyl function of one amino acid and the amino group of another amino acid are linked as follows:

$$R_1—COOH + R_2—NH_2 \rightarrow R_1—CO—NH—R_2 + H_2O.$$

Several hundred amino acids may be linked together by peptide bonds into long chains which assume specific configurations depending upon the nature, sequence, and number of amino acid constituents and the environmental medium. The concept of specific configuration of a protein means that it has a three-dimensional structure as well as a specific sequence and number of amino acid components. The biological function of the proteins is so tightly coupled with their specific configuration that any approach to the study of the biochemical mechanisms underlying the biological function must be considered in terms of the three-dimensional structure of these molecules.

Within the framework of a short monograph it is impossible to discuss the biochemistry of all of the various functions of proteins. Therefore protein structure in general is described as the basis for the understanding of the manifold biochemical activities of these macromolecules encountered by the physician and the biologist. Although the amino acids are interesting and important compounds in their own right, space limitation permits these compounds to be discussed solely in relation to their role in the structure and function of the protein molecule.

AMINO ACIDS

General Characteristics of Amino Acids

Amino acids have at least two functional groups, a carboxyl and an amino group (proline and hydroxyproline have an imino instead of an amino group). In the case of the amino acids found in proteins, an amino group is always attached to the carbon atom that is at the α position in relation to a carboxyl group (in the case of proline and hydroxyproline it is the imino group that is in the α position). The following is a general formula representing these α-amino acids:

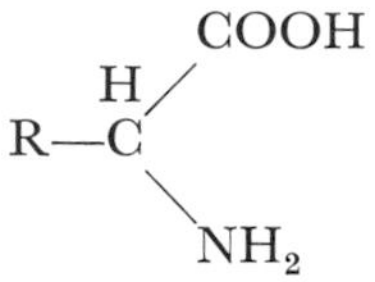

$$R—C \overset{\displaystyle H}{\underset{\displaystyle NH_2}{\overset{\diagup}{\diagdown}}} COOH$$

The R group is different for each amino acid.

Many amino acids possess a third functional group in addition to the carboxyl and its α-amino group; e.g., the R group may contain a sulf-hydryl or amino or carboxyl or phenolic or other functional group. Amino acids may exist in any one of the following forms, depending on the pH of the medium.

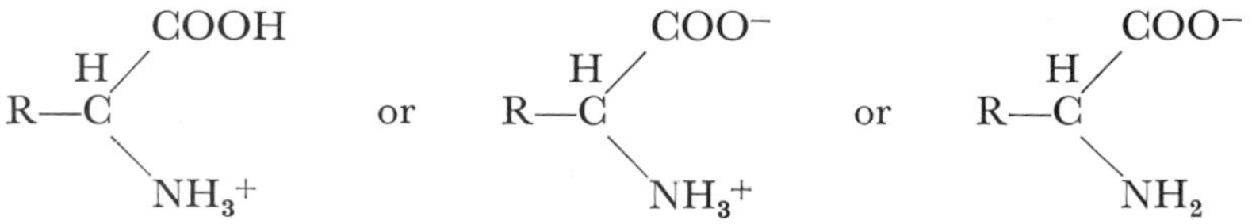

$$R—C \overset{\displaystyle H}{\underset{\displaystyle NH_3^+}{\overset{\diagup}{\diagdown}}} COOH \quad or \quad R—C \overset{\displaystyle H}{\underset{\displaystyle NH_3^+}{\overset{\diagup}{\diagdown}}} COO^- \quad or \quad R—C \overset{\displaystyle H}{\underset{\displaystyle NH_2}{\overset{\diagup}{\diagdown}}} COO^-$$

That is, amino acids are able to display either acidic or basic characteristics.

Amino Acids as Ampholytes

Figure 1-1 shows schematic titration curves obtained with different amino acids, with strong acid and alternatively strong alkali used as the titrant. Obviously amino acids behave as weak acids when a strong base is added as the titrant and as weak bases when strong acid is employed as the titrant.

At pH 7 the amino and carboxyl functions of the various amino acids shown in Figure 1-1 are almost completely ionized. This form may be

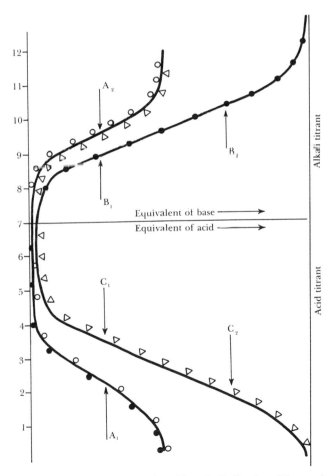

Figure 1-1. Titration of amino acids with acid and alkali. ○ = Monoaminomonocarboxylic acid. △ = Monoaminodicarboxylic acid. ● = Diaminomonocarboxylic acid. Arrows indicate inflection points.

represented by the following formulas:

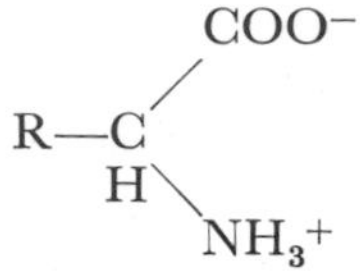

$$R-\underset{\underset{H}{|}}{C}\underset{NH_3^+}{\overset{COO^-}{<}}$$ Monoaminomonocarboxylic acid

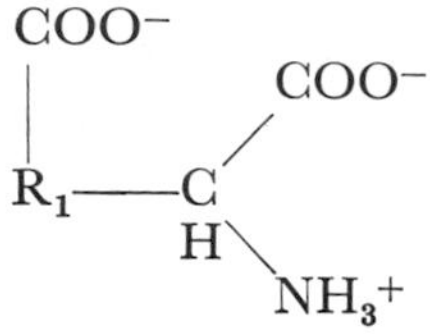

$$\underset{\underset{H}{|}}{\overset{COO^-}{\underset{|}{R_1-C}}}\underset{NH_3^+}{\overset{COO^-}{<}}$$ Monoaminodicarboxylic acid

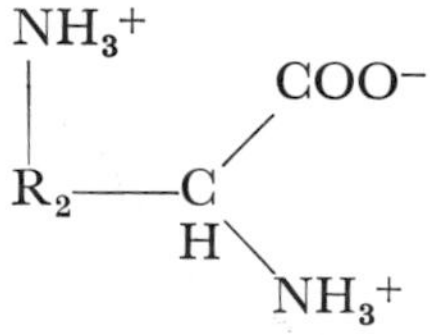

$$\underset{\underset{H}{|}}{\overset{NH_3^+}{\underset{|}{R_2-C}}}\underset{NH_3^+}{\overset{COO^-}{<}}$$ Diaminomonocarboxylic acid

At the end point of the alkali titration (pH $\sim$ 12) the monoaminomono-carboxylic acid and the monoaminodicarboxylic acid donate one proton and the diaminomonocarboxylic acid releases two protons to the titrant base. At the same time these amino acids are transformed into their respective alkaline structures. This form of the amino acids may be represented by the following formulas:

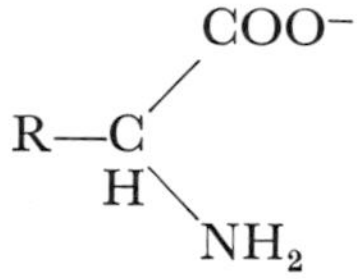

$$R-\underset{\underset{H}{|}}{C}\underset{NH_2}{\overset{COO^-}{<}}$$ Monoaminomonocarboxylic acid

$$\underset{\underset{H}{|}}{\overset{COO^-}{\underset{|}{R_1-C}}}\underset{NH_2}{\overset{COO^-}{<}}$$ Monoaminodicarboxylic acid

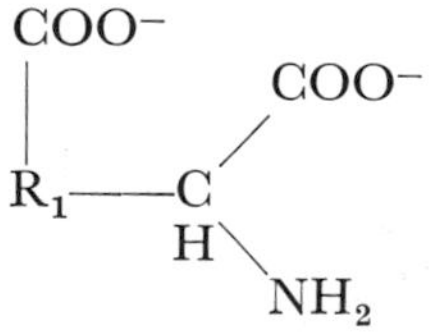

$$\underset{\underset{H}{|}}{\overset{NH_2}{\underset{|}{R_2-C}}}\underset{NH_2}{\overset{COO^-}{<}}$$ Diaminomonocarboxylic acid

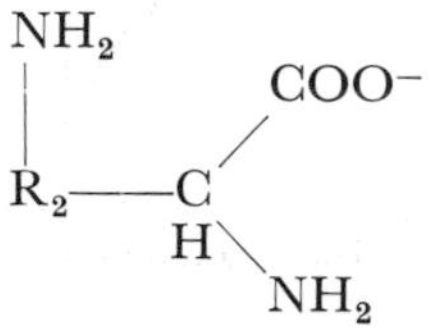

During the acid titration the isoelectric form of the amino acid changes to the acidic form by accepting protons from the donating titrant. The acidic forms may be represented by the following formulas:

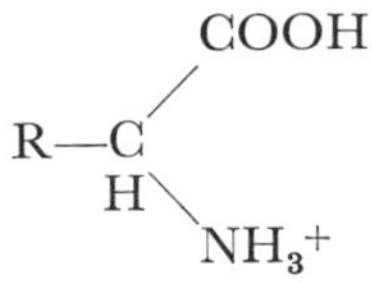

$$R-\overset{\displaystyle H}{\underset{}{C}}\overset{\displaystyle COOH}{\underset{\displaystyle NH_3^+}{}}$$ Monoaminomonocarboxylic acid

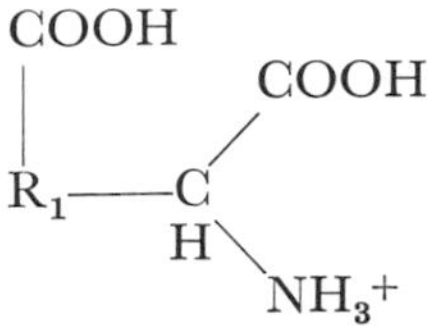

$$R_1-\overset{\displaystyle COOH}{\underset{\displaystyle H}{C}}\overset{\displaystyle COOH}{\underset{\displaystyle NH_3^+}{}}$$ Monoaminodicarboxylic acid

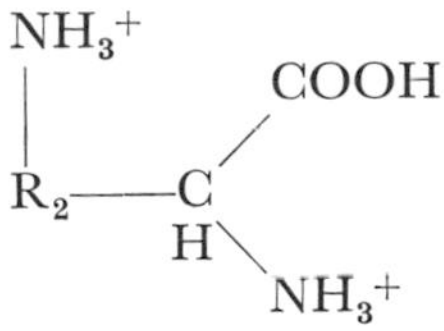

$$R_2-\overset{\displaystyle NH_3^+}{\underset{\displaystyle H}{C}}\overset{\displaystyle COOH}{\underset{\displaystyle NH_3^+}{}}$$ Diaminomonocarboxylic acid

The titration curve of the monoaminomonocarboxylic acid exhibits a single inflection point around pH 2.5 (arrow A_1 in Fig. 1-1) during acid titration and another inflection point during alkali titration at approximately pH 9.7 (arrow A_2). At around pH 2.5 approximately 50 per cent of the total amino acid is present in the isoionic form and 50 per cent in the acidic form, and at pH 9.7 approximately 50 per cent of the amino acid is present in the isoionic and 50 per cent in the basic form.

Monoaminodicarboxylic acids have two such inflection points in the acidic range because each carboxyl group takes up a proton at a somewhat different pH range. There is also a third inflection point in the alkaline range because of the proton release from the amino group (arrows C_1, C_2, and A_2).

The titration curve of diaminomonocarboxylic acids has one inflection point in the acidic but two in the alkaline range (arrows A_1, B_1, B_2 in Fig. 1-1); the α-amino group has a somewhat different proton attraction than the other amino group of the molecule.

Proteins, like their amino acid components, behave as ampholytes. The physical-chemical basis for many of the biological functions of proteins (e.g., buffering capacity, interactions between proteins and other charged molecules) related to this property. Therefore before considering the functioning of proteins, it is profitable to look somewhat more quantitatively into the ampholytic behavior of the amino acids.

Taking the simplest example, the titration of the monoaminomonocarboxylic acid, the following equilibrium equation may be written:

Acid Titration

Equation 1:

$$
\begin{bmatrix} R-C \begin{smallmatrix} COO^- \\ H \\ NH_3{}^+ \end{smallmatrix} \end{bmatrix} + H^+ \rightleftharpoons \begin{bmatrix} R-C \begin{smallmatrix} COOH \\ H \\ NH_3{}^+ \end{smallmatrix} \end{bmatrix}
$$

Equation 2:

$$
K_{a1} = \frac{\begin{bmatrix} R-C \begin{smallmatrix} COO^- \\ H \\ NH_3{}^+ \end{smallmatrix} \end{bmatrix}[H^+]}{\begin{bmatrix} R-C \begin{smallmatrix} COOH \\ H \\ NH_3{}^+ \end{smallmatrix} \end{bmatrix}}
$$

After rearranging:

Equation 3:

$$
[H^+] = K_{a1} \frac{\begin{bmatrix} R-C \begin{smallmatrix} COOH \\ H \\ NH_3{}^+ \end{smallmatrix} \end{bmatrix}}{\begin{bmatrix} R-C \begin{smallmatrix} COO^- \\ \\ NH_3{}^+ \end{smallmatrix} \end{bmatrix}}
$$

Expressing $[H^+]$ and K_a in terms of their respective negative logarithms (pH, pK) we obtain:

Equation 4:*

$$
pH = pK_{a1} + \log \frac{\begin{bmatrix} R-C \begin{smallmatrix} COO^- \\ \\ NH_3{}^+ \end{smallmatrix} \end{bmatrix}}{\begin{bmatrix} R-C \begin{smallmatrix} COOH \\ \\ NH_3{}^+ \end{smallmatrix} \end{bmatrix}}
$$

It is clear that the numerator of the log term of this equation differs from the denominator chemically in one respect only; namely, the carboxyl group is in the ionized form in the numerator and nonionized form in the

---

* For more complete discussion and general application of the Henderson-Hasselbalch equation, see references 1 and 2 in Chapter 2.

denominator. The numerator represents the fully ionized form of the amino acid, the concentration of which is decreasing as acid titration progresses. The denominator represents the acidic form of the amino acid, the concentration of which increases during the titration. At one pH during the titration, the concentration of each of the two species is the same, i.e., the inflection point of the titration curve. At this point the numerical value of the second right-hand term of equation 4 is 0 (i.e., the log of 1). Obviously, therefore, the pK_a of the carboxyl group is equal to that pH at which the number of ionized and nonionized amino acid carboxyl groups is equal.

The different forms of a monoaminomonocarboxylic acid during alkaline titration can be quantitatively considered in a similar manner.

$$\text{Equation 5:} \qquad \text{pH} = \text{p}K_{a2} + \log \frac{\left[\begin{array}{c} \text{COO}^- \\ R-C \\ H \\ \text{NH}_2 \end{array}\right]}{\left[\begin{array}{c} \text{COO}^- \\ R-C \\ H \\ \text{NH}_3^+ \end{array}\right]}$$

Here the denominator of the second right-hand term represents the ionized form of the amino group and the numerator the nonionized form. When these two species are present in equal concentrations the numerical value of this term is 0. Clearly the pK_{a2} has the same general meaning as pK_{a1}; specifically, it is equal to that pH at which the ionized and nonionized forms of the amino group are present in equal concentrations. This point is shown on the titration curve in Figure 1-1 by arrow A_2.

In proteins most α-amino and neighboring α-carboxyl groups of the amino acid residues are present as part of the covalent peptide links between the amino acids of the polypeptide chain and therefore are not free to ionize. On the other hand, the carboxyl and amino groups of the side chains are often free to donate or accept protons. Some amino acids contain groups other than the amino or carboxyl which may donate or accept protons, and when such groups are not involved in covalent bond formation within the protein structure, they function as proton donor and acceptor components of the proteins.

Table 1-1 shows the ionizable side chain groups of proteins. It may be seen that the pK_a of a given group in the free amino acid is somewhat different from the pK_a when the group is in the structure of a protein. These differences in the proton-donating or proton-accepting ability of a given kind of ionizable group are caused by the influence of neighboring amino acid residues and other components of the protein.

Table 1-1. *Ionizable Side Chain Groups of Proteins*

AMINO ACID	GROUP	IONIZATION	pK_a AMINO ACID	pK_a BOVINE SERUM ALBUMIN	pK_a (RANGE) IN INSULIN β-LACTOGLOBULIN OVALBUMIN LYSOZYME RIBONUCLEASE
Glutamic acid	γ-carboxyl	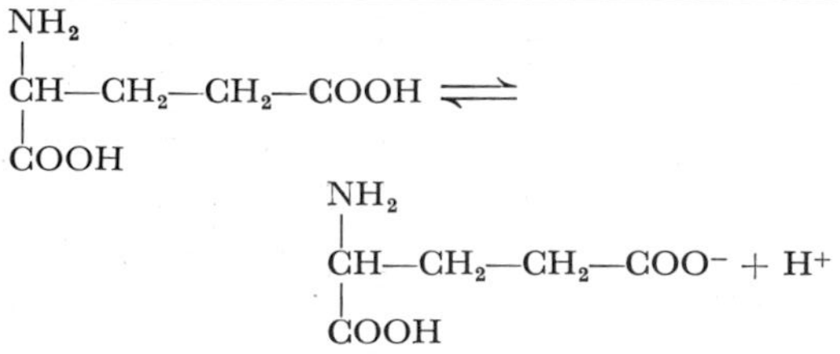 $\overset{NH_2}{\underset{COOH}{CH}}-CH_2-CH_2-COOH \rightleftharpoons \overset{NH_2}{\underset{COOH}{CH}}-CH_2-CH_2-COO^- + H^+$	3.91	3.95	4.3–4.7
Aspartic acid	β-carboxyl	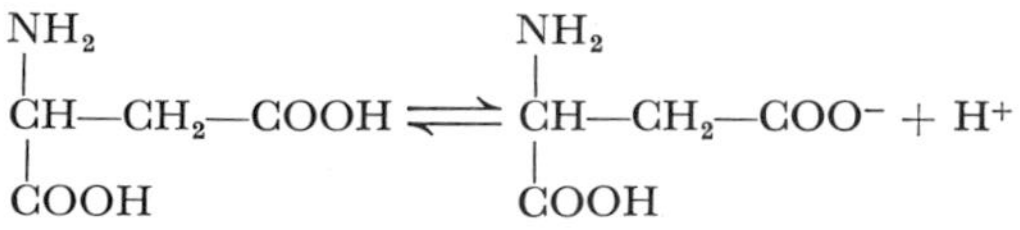 $\overset{NH_2}{\underset{COOH}{CH}}-CH_2-COOH \rightleftharpoons \overset{NH_2}{\underset{COOH}{CH}}-CH_2-COO^- + H^+$	4.25	3.95	4.3–4.7
Histidine	imidazole	$\underset{CH_2}{\overset{H^+N-CH}{\underset{\|}{HC=\!=\!CH-NH}}} \rightleftharpoons \underset{CH_2}{\overset{HC-N}{\underset{\|}{HC=\!=\!CH-NH}}} + H^+$ ($H_2N-\overset{H}{\underset{C}{|}}-COOH$)	6.00	7.00	6.4–7.0
Cysteine	sulfhydryl	$\underset{H_2N-\overset{H}{\underset{\|}{C}}-COOH}{CH_2-SH} \rightleftharpoons \underset{H_2N-\overset{H}{\underset{\|}{C}}-COOH}{CH_2-S^-} + H^+$	8.33	?	?

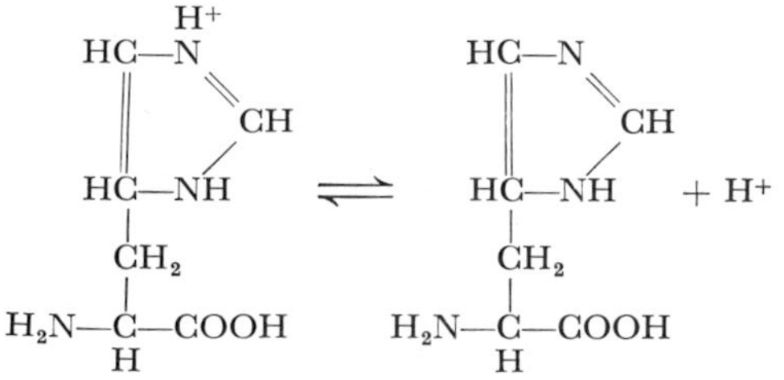

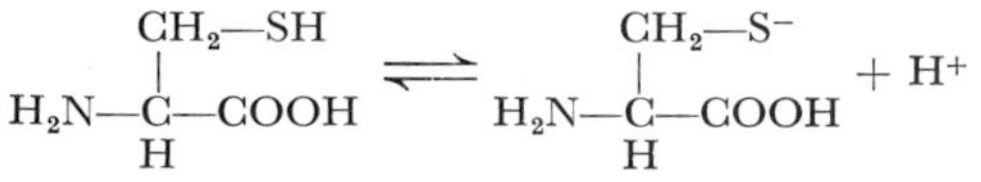

Lysine	ε-amino	$\begin{array}{c} NH_2 \\ | \\ CH-CH_2-CH_2-CH_2-CH_2-NH_2 \\ | \\ COOH \end{array} \overset{H^+}{\rightleftharpoons} \begin{array}{c} NH_2 \\ | \\ C-CH_2-CH_2-CH_2-CH_2-NH_2 + H^+ \\ | \\ COOH \end{array}$	10.53	9.80	10.1–10.6
Tyrosine	phenolic hydroxyl	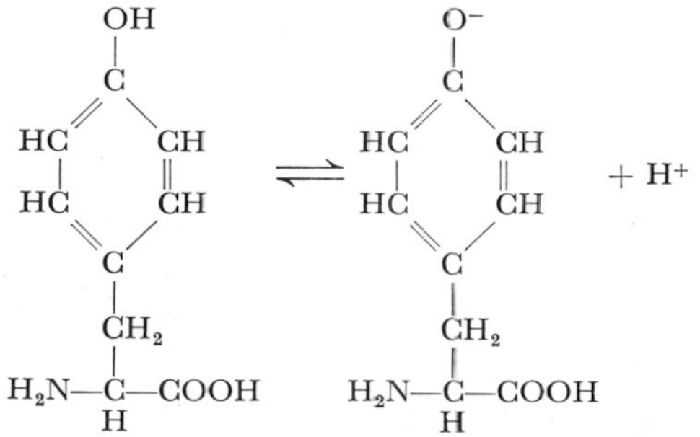 $HO-C\begin{array}{c}CH=CH\\ \\ HC=CH\end{array}C-CH_2-\underset{H}{\overset{H_2N}{C}}-COOH \rightleftharpoons O^--C\begin{array}{c}CH=CH\\ \\ HC=CH\end{array}C-CH_2-\underset{H}{\overset{H_2N}{C}}-COOH + H^+$	10.07	10.35	8.5–10.9
Arginine	guanidino	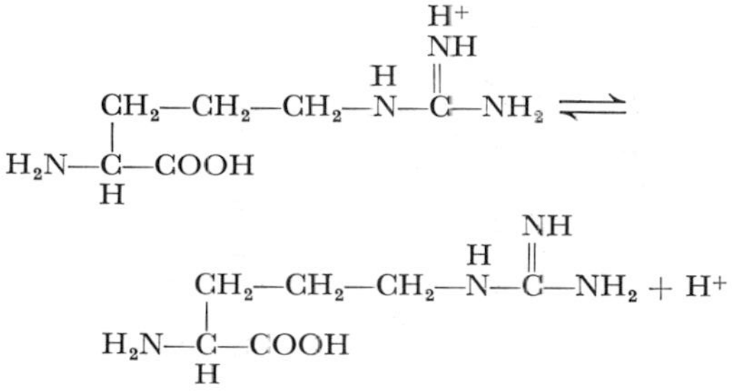 $H_2N-\underset{H}{\overset{}{C}}-COOH\;CH_2-CH_2-CH_2-N-\underset{\overset{+}{NH_2}}{\overset{H}{\underset{}{C}}}-NH_2 \rightleftharpoons CH_2-CH_2-CH_2-N-\underset{NH}{\overset{H}{\underset{}{C}}}=C-NH_2 + H^+ \;\; H_2N-\underset{H}{\overset{}{C}}-COOH$	12.48	>12.0	11.9–13.3

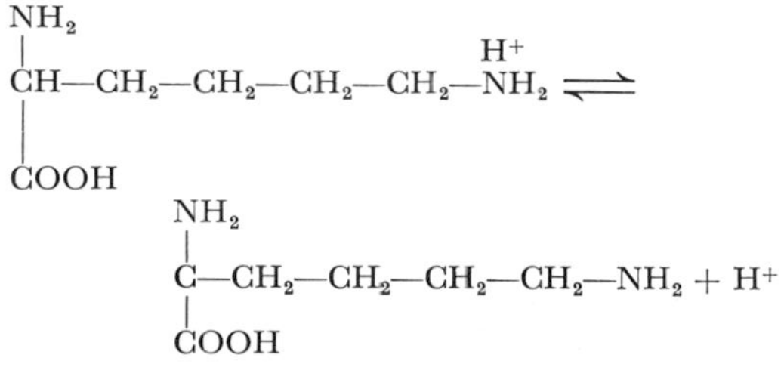

Optical Rotation of Amino Acids

Another characteristic of all amino acids, except glycine, is the asymmetrical character of the α carbon; i.e., the α carbon has four different substituents.

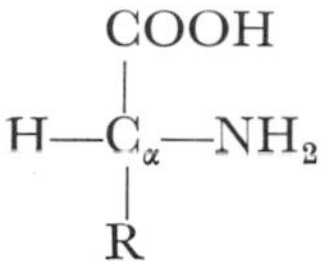

$$
\begin{array}{c}
COOH \\
| \\
H-C_\alpha-NH_2 \\
| \\
R
\end{array}
$$

This asymmetrical carbon atom is responsible for optical activity of these amino acids which is measured by the degree to which a beam of polarized light is rotated as it passes through a solution of the optically active compound. Most naturally occurring amino acids rotate polarized light in a counterclockwise direction and, therefore, are designated as levorotatory or L-amino acids. Some dextrorotatory or D-amino acids have also been isolated from natural products.

Classification of Amino Acids Found in Protein Structure

Classification of amino acids may be attempted in several different ways. In this text the acid-base characteristic of the amino acid in the fully ionized form will serve as the basis of classification. (The structural formulas given here do not indicate ionization; this has been discussed in the preceding section.)

Neutral Amino Acids

Glycine: H_2N-CH_2-COOH

Alanine: $H_2N-CH-COOH$
$$\begin{array}{c}|\\CH_3\end{array}$$

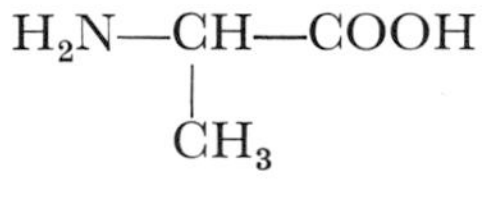

Valine: $H_2N-CH-COOH$
$$\begin{array}{c}|\\CH\\ \diagup \diagdown \\ CH_3 \quad CH_3\end{array}$$

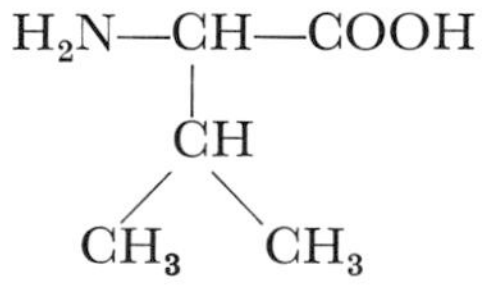

Leucine: $H_2N-CH-COOH$
$$\begin{array}{c}|\\CH_2\\|\\CH\\ \diagup \diagdown \\ CH_3 \quad CH_3\end{array}$$

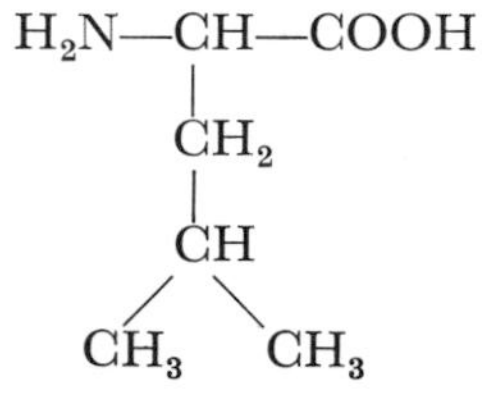

Isoleucine:

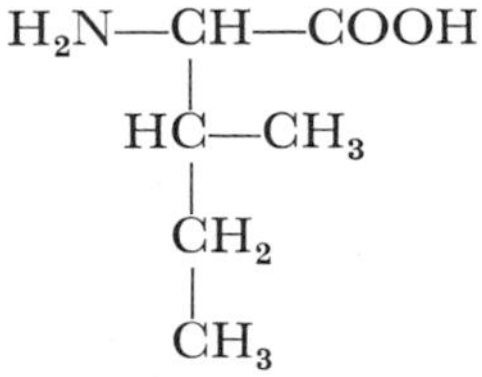

$$H_2N\text{—}CH\text{—}COOH$$
$$|$$
$$HC\text{—}CH_3$$
$$|$$
$$CH_2$$
$$|$$
$$CH_3$$

Glycine, alanine, valine, leucine, and isoleucine differ from each other mainly in regard to the length and branching of the hydrocarbon side chain.

Methionine:

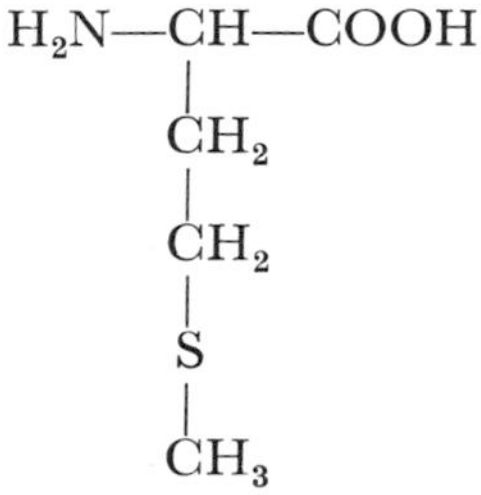

$$H_2N\text{—}CH\text{—}COOH$$
$$|$$
$$CH_2$$
$$|$$
$$CH_2$$
$$|$$
$$S$$
$$|$$
$$CH_3$$

Methionine contains a thioether group in addition to the hydrocarbon side chain.

Phenylalanine:

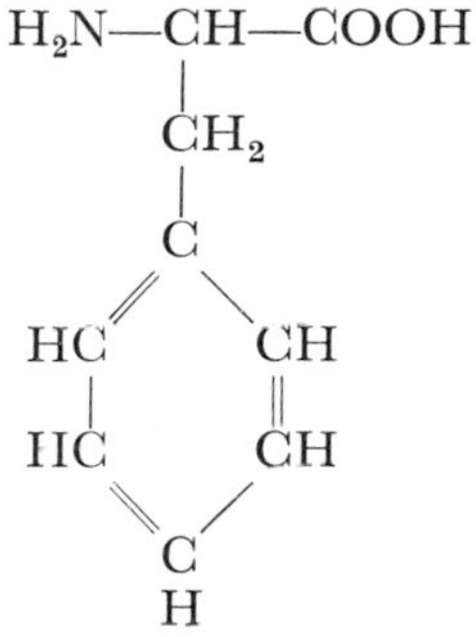

$$H_2N\text{—}CH\text{—}COOH$$
$$|$$
$$CH_2$$
$$|$$
$$C$$

$$HC \quad\quad CH$$
$$HC \quad\quad CH$$
$$C$$
$$H$$

Proline:

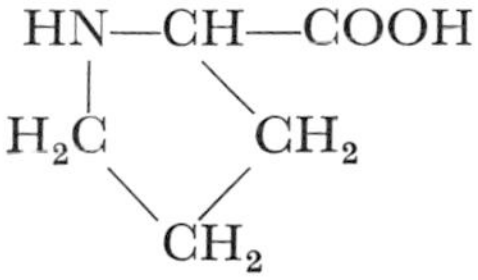

$$HN\text{—}CH\text{—}COOH$$
$$H_2C \quad\quad CH_2$$
$$CH_2$$

Hydroxyproline:

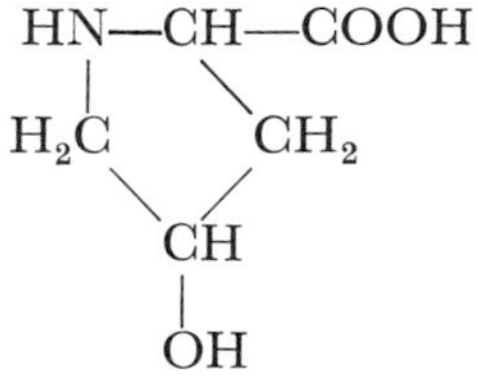

$$HN\text{—}CH\text{—}COOH$$
$$H_2C \quad\quad CH_2$$
$$CH$$
$$|$$
$$OH$$

Phenylalanine contains an aromatic ring, whereas proline and hydroxyproline are heterocyclic ring compounds containing an imino nitrogen. Hydroxyproline is found almost exclusively in collagen.

Serine:

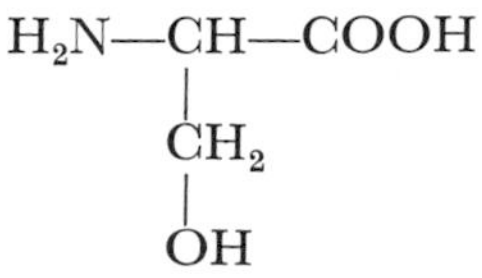

$$H_2N\text{—}CH\text{—}COOH$$
$$|$$
$$CH_2$$
$$|$$
$$OH$$

Threonine:

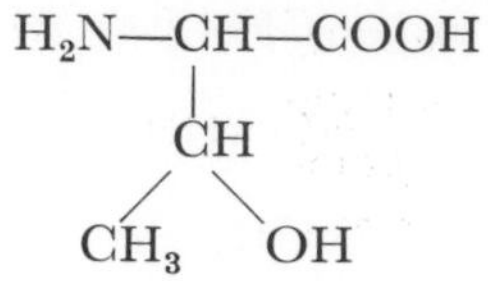

$$H_2N—CH—COOH$$
$$|$$
$$CH$$
$$CH_3 \quad OH$$

Serine and threonine contain a primary and secondary alcohol group, respectively.

Tryptophan:

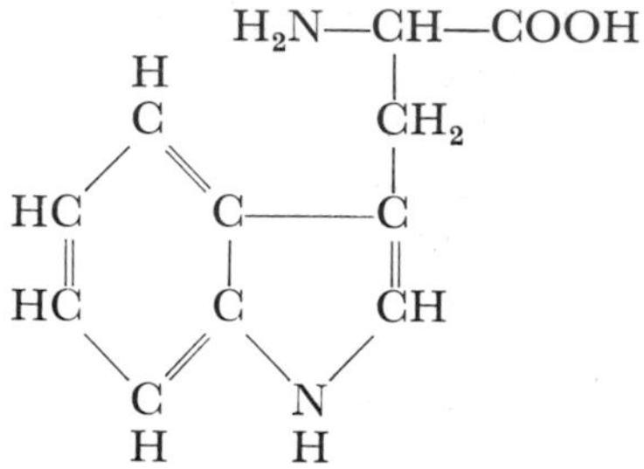

$$H_2N—CH—COOH$$

Tryptophan is an amino acid with a nitrogen-containing aromatic ring structure.

Asparagine:

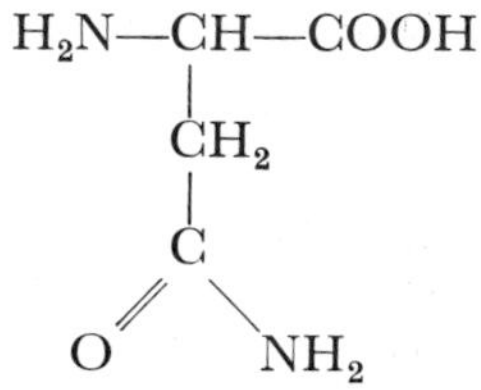

$$H_2N—CH—COOH$$
$$|$$
$$CH_2$$
$$|$$
$$C$$
$$O \quad NH_2$$

Glutamine:

$$H_2N—CH—COOH$$
$$|$$
$$CH_2$$
$$|$$
$$CH_2$$
$$|$$
$$C$$
$$O \quad NH_2$$

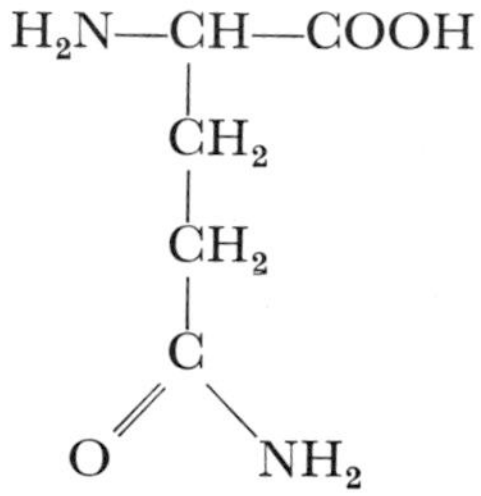

Asparagine and glutamine are the acid amides of aspartic and glutamic acids.

Cystine:

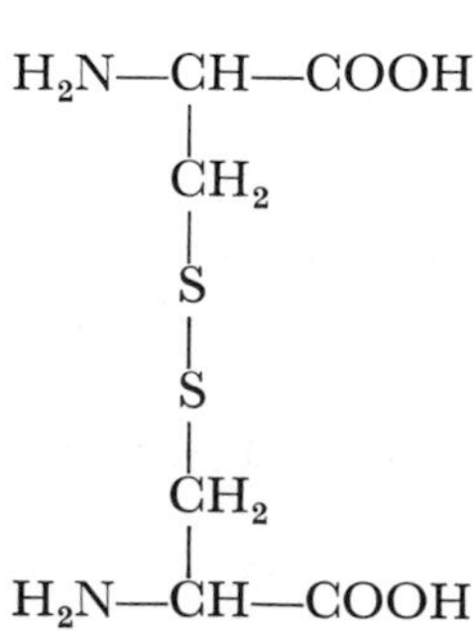

$$H_2N—CH—COOH$$
$$|$$
$$CH_2$$
$$|$$
$$S$$
$$|$$
$$S$$
$$|$$
$$CH_2$$
$$H_2N—CH—COOH$$

Cystine is an amino acid containing two carboxyl groups adjacent to α-amino groups with a disulfide bridge linking the structures into a single unit.

Acidic Amino Acids

Aspartic acid:

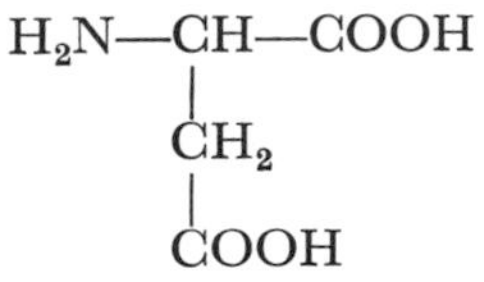

$$H_2N-CH-COOH$$
$$|$$
$$CH_2$$
$$|$$
$$COOH$$

Glutamic acid:

$$H_2N-CH-COOH$$
$$|$$
$$CH_2$$
$$|$$
$$CH_2$$
$$|$$
$$COOH$$

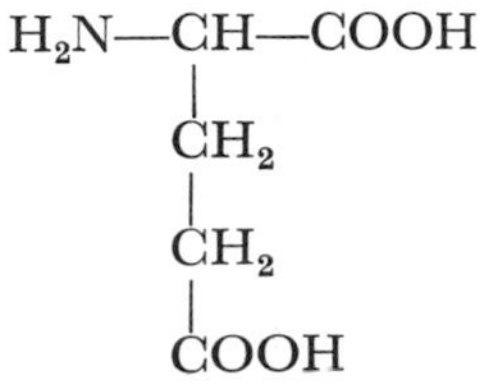

Aspartic and glutamic acids are dicarboxylic acids and in the neutral pH range both of the carboxyl groups of these compounds are ionized. Therefore in the physiological pH range they have a net negative charge.

Cysteine:

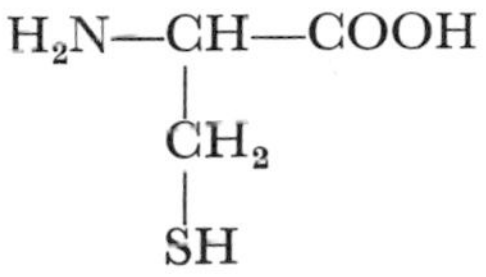

$$H_2N-CH-COOH$$
$$|$$
$$CH_2$$
$$|$$
$$SH$$

Cysteine is a sulfhydryl-containing amino acid; such thiol groups are among the most reactive groups in a protein molecule.

Tyrosine:

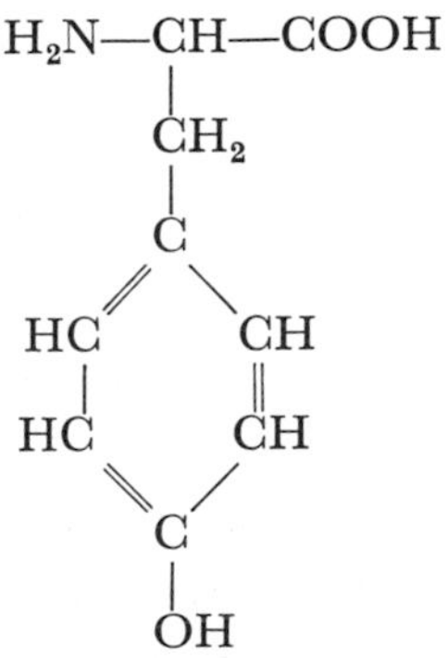

$$H_2N-CH-COOH$$
$$|$$
$$CH_2$$
$$|$$
$$C$$

$$HC\quad\quad CH$$
$$HC\quad\quad CH$$

$$C$$
$$|$$
$$OH$$

Tyrosine is a phenol-containing amino acid.

Basic Amino Acids

Lysine:

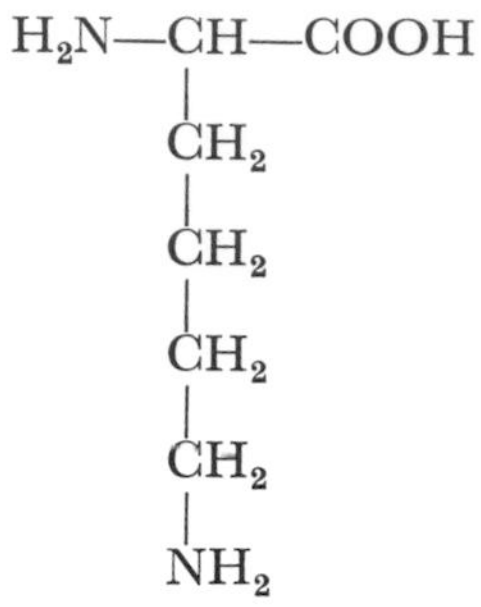

$$H_2N—CH—COOH$$
$$|$$
$$CH_2$$
$$|$$
$$CH_2$$
$$|$$
$$CH_2$$
$$|$$
$$CH_2$$
$$|$$
$$NH_2$$

Arginine:

$$H_2N—CH—COOH$$
$$|$$
$$CH_2$$
$$|$$
$$CH_2$$
$$|$$
$$CH_2$$
$$|$$
$$NH$$
$$|$$
$$C$$
$$H_2N \diagup \quad \diagdown\!\!\!= NH$$

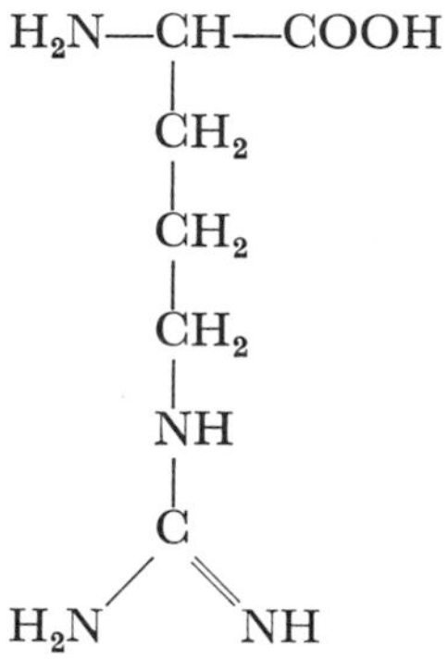

Both arginine and lysine carry a net positive charge in the physiological pH range.

Histidine:

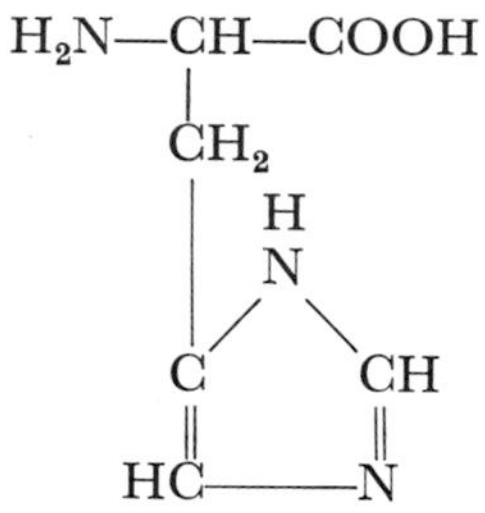

$$H_2N—CH—COOH$$
$$|$$
$$CH_2$$

The imidazole ring structure of histidine is a very reactive group.

Biologically Important Amino Acids Not Found in Protein Structure

In addition to the amino acids participating in protein structure, there are a number of biologically important amino acids and amino acid derivatives that are not found in proteins. Some of these substances are:

β-Alanine: $H_2N—CH_2—CH_2—COOH$

β-Alanine is a constituent of the peptides carnosine and anserine, compounds present in large quantities in the muscle, and it is also a component of coenzyme A.

γ-Aminobutyric acid:

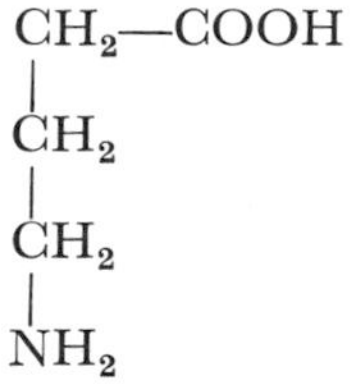

$$CH_2-COOH$$
$$|$$
$$CH_2$$
$$|$$
$$CH_2$$
$$|$$
$$NH_2$$

This compound is found in the brain tissue of mammals and is also present in some bacteria and plants.

Sarcosine:

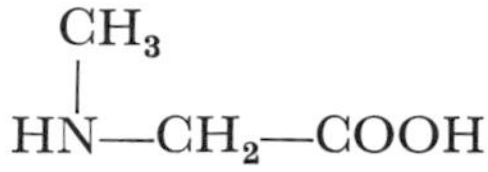

$$CH_3$$
$$|$$
$$HN-CH_2-COOH$$

Sarcosine is an intermediate in the one-carbon compound metabolism.

Histamine:

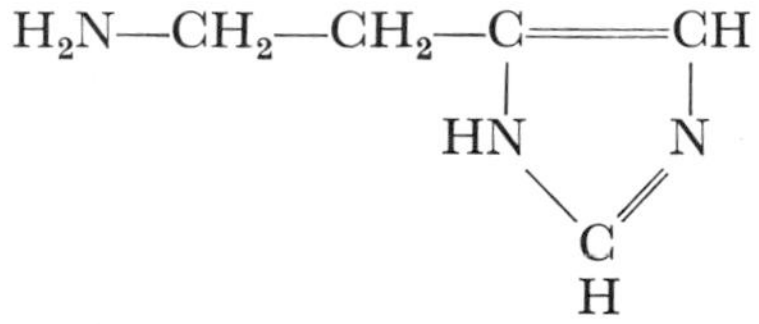

$$H_2N-CH_2-CH_2-C{=\!=}CH$$

with ring HN, N, C, H

Histamine is a powerful vasodilator which may be produced in mammalian organisms by the decarboxylation of histidine. There are at least two different enzymes, aromatic L-amino acid decarboxylase and histidine decarboxylase, which catalyze this process.

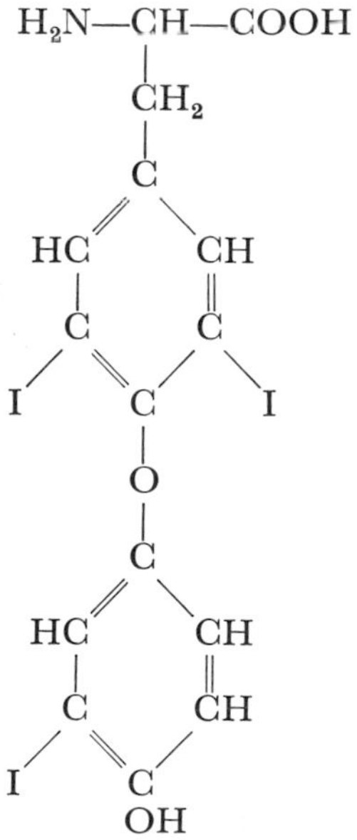

$$H_2N-CH-COOH$$

Triiodothyronine (T₃)

$$H_2N-CH-COOH$$

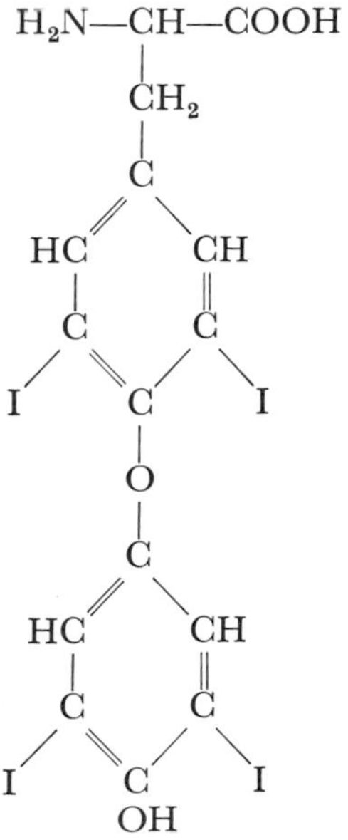

Thyroxine (T₄)

The amino acids thyroxine and triiodothyronine are important hormones of the thyroid gland which are involved in the regulation of cellular metabolism. They are iodinated derivatives of thyroxine.

Epinephrine:

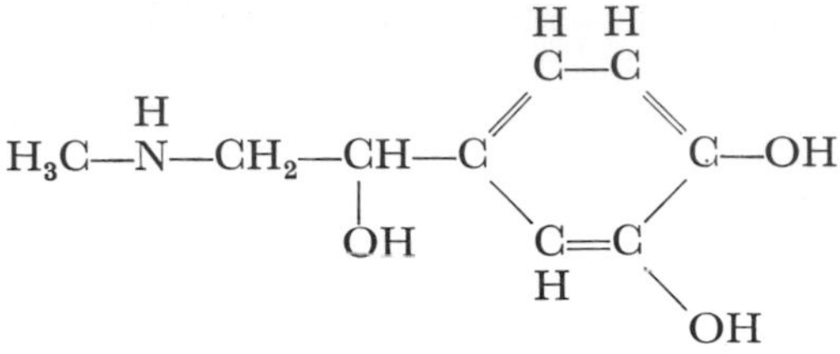

$$H_3C-\overset{\underset{\displaystyle |}{H}}{N}-CH_2-CH-C \quad\quad\quad C-OH$$

Epinephrine and the closely related norepinephrine are tyrosine derivatives which function importantly as hormones and as neurohumors.

BIOLOGICALLY IMPORTANT PEPTIDES

Almost every tissue of the mammalian organism contains peptides, examples of which are presented here:

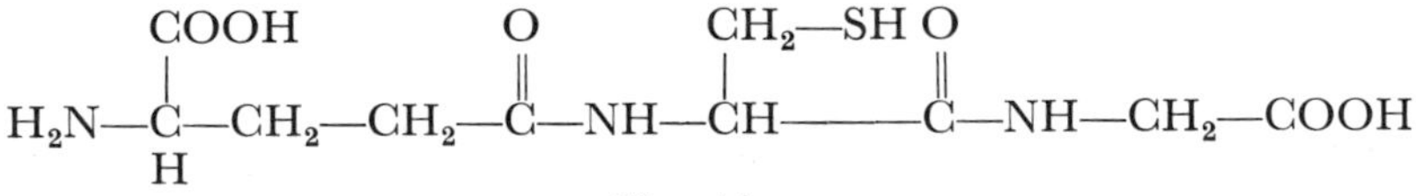

$$H_2N-\overset{\underset{\displaystyle |}{\underset{\displaystyle H}{}}}{\overset{\displaystyle COOH}{C}}-CH_2-CH_2-\overset{\underset{\displaystyle O}{\parallel}}{C}-NH-\overset{\underset{\displaystyle |}{\overset{\displaystyle CH_2-SH}{}}}{CH}\overset{\underset{\displaystyle O}{\parallel}}{C}-NH-CH_2-COOH$$

Glutathione
(γ-L-Glutamyl-L-cysteinylglycine)

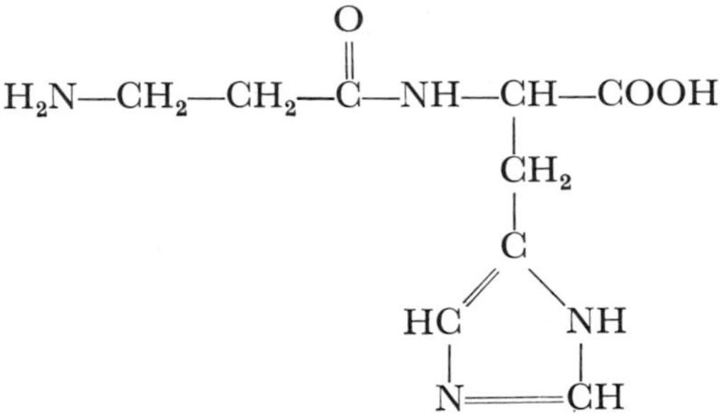

$$H_2N-CH_2-CH_2-\overset{\underset{\displaystyle O}{\parallel}}{C}-NH-CH-COOH$$

Carnosine (β-Alanylhistidine)

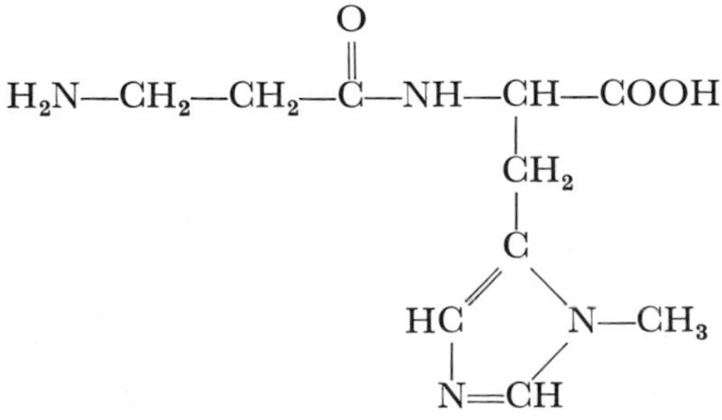

$$H_2N-CH_2-CH_2-\overset{\underset{\displaystyle O}{\parallel}}{C}-NH-CH-COOH$$

Anserine (β-Alanyl-1-methylhistidine)

Glutathione may function as a coenzyme; the biochemical role of carnosine and anserine is not known. A number of peptide hormones (see Chapter 8) have been isolated and their structures determined. Because of the biological importance of certain peptides, methods for the chemical synthesis of such compounds have been developed.

REFERENCES

1. Greenstein, J. P., and Winitz, M.: Chemistry of the Amino Acids. 3 vols. New York, John Wiley & Sons, 1961.
2. Kopple, K. D.: Peptides and Amino Acids. New York, W. A. Benjamin, Inc., 1966.
3. Meister, A.: Biochemistry of the Amino Acids. New York, Academic Press, 1965. (Chapter 1 in Volume 1 deals with the general chemical properties of amino acids and peptides.)

2

PROTEINS AS MACROMOLECULES

PHYSICOCHEMICAL CHARACTERIZATION OF MACROMOLECULES

In order to understand the biochemical mechanism of action of various macromolecules it is first necessary to characterize these molecules as physicochemical entities. Whenever a macromolecule is crystallizable, x-ray crystallography appears to be the best direct method to explore the structural characteristics of that molecule in the solid state. If the molecule is not crystallizable, application of x-ray diffraction is limited. In such cases methods for the physical-chemical characterization of macromolecules in solution must be used.

Some macromolecules contain similar subunits (e.g., DNA) but differ from each other with respect to molecular weight. In such cases the determined molecular weight is an average number, characteristic of the population of the molecules in solution rather than being true for each molecule. Depending upon the nature of the method used for the molecular weight determination, several different average values may be obtained, e.g., "weight average molecular weight" and the "number average molecular weight."

The "number average molecular weight" (Mn) may be obtained by osmotic pressure measurement or by molecular weight calculations based on chemical end group determination.

Equation 1: $$Mn = \frac{\text{Total weight of the material}}{\text{Total number of molecules}}$$

Weight average molecular weight (Mw) is obtained by light scattering and ultracentrifugation (equilibrium or sedimentation velocity together with diffusion and other measurements) and is defined as the sum of the molecular weights times their weights for all molecular species divided by the total weight of all molecular species present in the solution.

$$\text{Equation 2:} \quad \text{Mw} = \Sigma \left(\frac{\text{Molecular weight of fraction 1} \times \text{weight of fraction 1}}{\text{weight of total material}} \right)$$

It is evident that in the calculation of Mn the contribution of each molecule in the population to the result is the same, while in the Mw calculation the individual molecules do not influence the result equally, but rather, heavier molecules affect it much more than the lighter molecules. Therefore by measuring both Mw and Mn considerable insight can be gained on the molecular weight distribution within a given macromolecular solution. If the numerical values of Mw and Mn are identical, the macromolecules of the solution must be homogeneous in regard to molecular weight. Such is the case with most crystallized globular proteins.

Biologically important macromolecules may be composed of one or of several chains. The chains are often folded so that a stable configuration is ensured by the intrachain or interchain bonds, or both. Such molecules may have the hydrodynamic or optical characteristics of spheres or rigid rods. In such cases the shape and size of the molecule are estimated by appropriate physical measurements. Other macromolecules lack such rigid internal structure and behave in solution as random coils. In this latter case the physicochemical measurements still permit the estimation of the effective radius of the molecule or the volume which is occupied by it.

Another important parameter of macromolecular solution which may be determined by physicochemical measurement is the distribution of these molecules in solution. Some molecules (such as spherical proteins) often occupy the available space randomly without any overlapping, while other molecules (such as random coils or fibrous protein) appear to overlap each other, even in moderately dilute solution. In this latter case high dilution is necessary if a measurement is to yield results characteristic for the individual molecules rather than the interacting molecules.

Another fundamentally important characteristic of proteins and nucleic acids is the electrostatic charge that these macromolecules carry in solution. Molecular interactions are to a significant extent determined by these charges and therefore the assessment of sign, quantity, and chemical nature of such charges is required for an understanding of the chemical basis of their biological actions.

Osmotic Pressure Measurements and Macromolecular Structure

Osmotic pressure measurements of solutions of macromolecules provide considerable information regarding the structure of these compounds.

If a solution is separated from pure solvent by a membrane which is permeable to the solvent molecules but not to the solute molecules, then solvent molecules undergo net migration from solvent to solution through the membrane. This net migration of solvent molecules can be stopped by applying an appropriate amount of hydraulic pressure to the solution; the osmotic pressure of the solution is defined in terms of the hydraulic pressure necessary to stop this net transfer of solvent molecules. The van't Hoff equation relates osmotic pressure, concentration of solute, and molecular weight of solute as follows:

Equation 3: $$\Pi = \frac{C}{M} RT$$

Π = osmotic pressure

C = concentration of the solute

M = molecular weight of the solute

R = Renault constant

T = absolute temperature

Equation 3 can be rearranged as follows:

Equation 4: $$\frac{\Pi}{C} = \frac{1}{M} RT$$

According to equation 4, $\frac{\Pi}{C}$ is a constant when temperature is held constant and a pure compound serves as solute. In practice, however, $\frac{\Pi}{C}$ proves not to be constant, as can be seen from Figure 2-1. Only an ideal solution, not a real one, behaves in this way. Some real solutions behave as depicted by curve 1 in Figure 2-1; such a deviation from ideal indicates a strong interaction between solute and solvent, leading to the immobilization of large numbers of solvent molecules in the immediate vicinity of the solute. Fibrous proteins and solvents which have great solvating effects often show such behavior. In other words, this may be interpreted as large-volume filling capacity of the solute. When the solution behaves as in curve 3 of

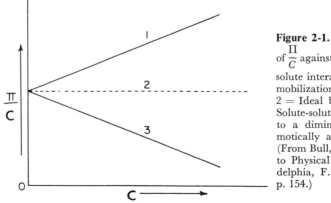

Figure 2-1. Diagrammatic plots of $\frac{\Pi}{C}$ against C. Curve 1 = Solvent-solute interaction leading to "immobilization" of solvent. Curve 2 = Ideal behavior. Curve 3 = Solute-solute interaction leading to a diminished number of osmotically active solute particles. (From Bull, H.: An Introduction to Physical Biochemistry. Philadelphia, F. A. Davis Co., 1964, p. 154.)

Figure 2-1, there is indication of strong solute-solute interactions such as an aggregation which diminishes the number of the osmotically active solute molecules.

Electrically charged macromolecules may bind small ions and influence their distribution between the solution and the solvent compartment, i.e., the Donnan effect, thus leading to an increase in the number of the osmotically active molecules in the solution. (The result is therefore falsely high osmotic pressure, which would indicate an erroneously low molecular weight.) This effect may be reduced to a minimum by carrying out the measurement in a medium containing relatively high salt concentration on both sides of the semipermeable membrane. Other problems in connection with osmotic pressure measurements have been the slowness of the procedure and the possibility of denaturation or aggregation occurring during the measurement itself. These have been eliminated to a great extent by the use of osmometers which measure the rate of solvent migration rather than the pressure directly.

Osmotic measurements obviously provide important information regarding the size and shape of macromolecules. In using this procedure for molecular weight determinations one must take into consideration the extent to which the solution is behaving in a nonideal fashion as well as contributions due to other factors such as the Donnan effect.

Diffusion and Macromolecular Structure

The diffusion characteristics of macromolecules provide information on the structure of these compounds. The following paragraphs briefly outline the principles underlying diffusion studies.

If solution is layered on the top of the solvent in a single phase, the solute molecules begin to move from the solution into the solvent, i.e., from the region of high solute molecule concentration to one of low solute concentration. At equilibrium the distribution of the solvent and solute molecules is equal throughout the system.

The Fick law permits a quantitative consideration of the diffusion phenomenon.

Equation 5:
$$\frac{ds}{dt} = -DA\,\frac{dc}{dx}$$

$\dfrac{ds}{dt}$ = mass of solute diffusing in t time

D = diffusion coefficient. This is expressed as the amount of material which diffuses across a 1 cm.^{2} surface in 1 sec. if the concentration gradient is also unity. Accordingly, if $A = 1$ cm.^{2}, $dt = 1$ sec., and $\dfrac{dc}{dx}$ = unity, then $-D = ds$

A = area available for diffusion

$\dfrac{dc}{dx}$ = concentration gradient along the distance (x) between solution and solvent

The diffusion coefficient is a function of the shape and size of the molecule and may be expressed in terms of a force (F) which is sufficient to provide 1 cm. per sec. velocity to the diffusing substance. Accordingly,

Equation 6:
$$F = \frac{RT}{D}$$

F = frictional coefficient as just defined

For the determination of D, the concentration of solute in the solution layer (c) and the area of the surface available for diffusion must be known. During the measurement the amount of solute (s) which diffused into the solvent at h distance from the solution in t time must be determined. From such data D may be calculated as follows:

Equation 7:
$$D = \frac{sh}{ctA}$$

In this equation h and A are instrumental constants, c is known by the investigator, and t has to be chosen in such a way as to allow s to be measured by chemical or optical methods.

Since D (the diffusion coefficient) depends primarily on the shape and size of the molecules, its value should be independent of the concentration of the solute if ideal solutions are studied. In practice such is not the case because the extent of solute-solute and solute-solvent interactions is influenced by solute concentrations and such interactions influence the numerical value of D.

Diffusion measurements are widely used to obtain information about the size and shape of protein molecules. Determination of the diffusion constant together with the sedimentation constant permits the calculation of the molecular weight of a protein. If the molecular weight of a protein is known, the determination of the diffusion constant may be used for calculations leading to information about the size and shape of the molecule. These physical parameters are most important for an understanding of the biological functions of these macromolecules.

Ultracentrifugation and Macromolecular Structure

The ultracentrifuge is an instrument capable of generating forces more than 100,000 times that of gravity. In the analytical ultracentrifuge there are optical devices which make it possible to observe the distribution of the macromolecules under the influence of the increased field force generated by the centrifugation (Fig. 2-2). This increased field force imposes a driving force upon the molecules which were equally distributed at the beginning of the experiment and this force leads to the appearance of concentration gradients.

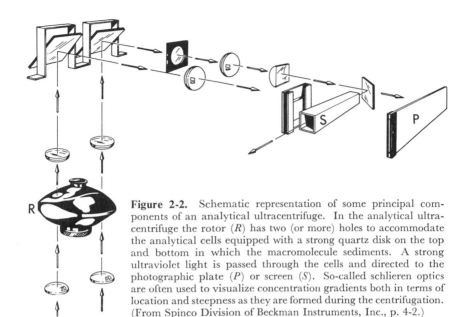

Figure 2-2. Schematic representation of some principal components of an analytical ultracentrifuge. In the analytical ultracentrifuge the rotor (*R*) has two (or more) holes to accommodate the analytical cells equipped with a strong quartz disk on the top and bottom in which the macromolecule sediments. A strong ultraviolet light is passed through the cells and directed to the photographic plate (*P*) or screen (*S*). So-called schlieren optics are often used to visualize concentration gradients both in terms of location and steepness as they are formed during the centrifugation. (From Spinco Division of Beckman Instruments, Inc., p. 4-2.)

The driving force acting on a molecule at distance *x* from the center of rotation (Fig. 2-3) can be quantitatively expressed as follows:

Equation 8:   Driving force $= M\omega^2 x(1 - \bar{v}\rho)$

M = molecular weight
ω = angular velocity of the rotor in radians/sec.
$\bar{v}$ = partial specific volume of the macromolecule (for most proteins, 0.75)
ρ = density of solute

As the macromolecule begins to accelerate because of the increased gravitational field force, it meets with resisting forces due to the viscous resistance of the solution.

Equation 9:   Resisting force $= F\dfrac{dx}{dt}$

$\dfrac{dx}{dt}$ = velocity of the sedimenting molecule

F = frictional coefficient

At sufficiently high sedimentation velocity the resisting force is just sufficient to balance the driving force and therefore,

Equation 10: $M\omega^2 x(1 - \bar{v}\rho) = F\dfrac{dx}{dt}$

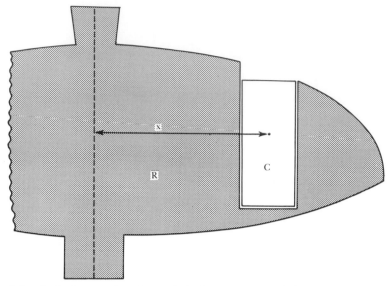

Figure 2-3.   Sagittal transection of an analytical rotor. R = Analytical rotor. C = Analytical cell. x = Distance between the center of rotation and a sedimenting molecule.

Since F was previously defined as $F = \dfrac{RT}{D}$, the equation may be rewritten as follows:

Equation 11: $$M\omega^2 x(1 - \bar{v}\rho) = \frac{RT}{D} \cdot \frac{dx}{dt}$$

The equation may be solved for M.

Equation 12: $$M = \frac{RT}{D(1 - \bar{v}\rho)} \cdot \frac{\dfrac{dx}{dt}}{x\omega^2}$$

and if

Equation 13: $$\frac{\dfrac{dx}{dt}}{x\omega^2} = S \text{ (Svedberg coefficient or sedimentation coefficient)}$$

then

Equation 14: $$M = \frac{RTS}{D(1 - \bar{v}\rho)}$$

With this approach two different measurements are needed to determine the molecular weight of a macromolecule, the determination of the diffusion constant (as discussed before) and the determination of the sedimentation constant. The sedimentation coefficient as defined in equation 14 depends primarily on the weight, size, and shape of the molecule. Under ideal conditions, when no solute-solute or solute-solvent interactions occur, the sedimentation coefficient is not affected by the concentration of solute. However, nucleic acids and most proteins display marked solute-solvent or solute-solute interactions. Therefore, the sedimentation constant must be computed by measuring the sedimentation coefficients at various substrate concentrations and extrapolating to zero substrate concentration.

Another ultracentrifugation method, the so-called sedimentation equilibrium method, permits the calculation of molecular weight without determining the diffusion constant. In this case the centrifugation, which is carried out at a moderate rpm, must be continued until the redistribution of solids in the cell has reached a final and constant state, i.e., until no net transfer of material at any cross-section in the cell is occurring. At this point the tendency of the centrifugal field force to build a concentration gradient is exactly counterbalanced by the free diffusion resulting from the concentration gradient. When this stage has been reached, the molecular weight of the centrifuged material may be computed by the following equation:

Equation 15:
$$M = \frac{2RT \ln C_2/C_1}{\omega^2 (1 - \bar{v}\rho)(x_2^2 - x_1^2)}$$

(where C_2 and C_1 are the concentrations of the centrifuged macromolecule at distance x_2 and x_1, respectively). A drawback of this method is the long centrifugation time needed to reach the equilibrium with large macromolecules, but recently the introduction of shorter cells has circumvented this problem.

Viscosity Measurements and Macromolecular Structure

Viscosity measurements are often used as an adjuvant method to determine size and shape of macromolecules. Since the apparatus required is not expensive and the technique is simple, this should be a preferred approach but unfortunately the theoretical interpretation of the viscosity measurements is complicated. Newton described viscosity as lack of slipperiness; i.e., it relates to the internal resistance of the relative motion of the molecules of one liquid layer with respect to molecules of the adjacent layer. This "internal resistance" is a function of the size and shape of the molecules of a liquid and the extent of solute-solute and solute-solvent interactions. According to Newton, the shearing forces (F) between two parallel layers of liquid in relative motion are related to the surface area of these layers (A), to the velocity gradient between the two moving layers $\left(\dfrac{dv}{dl}\right)$, and to some proportionality factor, called viscosity coefficient (η), characteristic of the molecules of these layers (Fig. 2-4). The quantitative relationship is expressed by the following equation:

Equation 16:
$$\eta = \frac{\dfrac{F}{A}}{\dfrac{dv}{dl}}$$

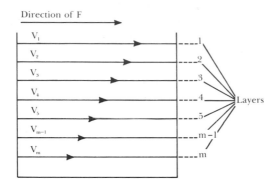

Figure 2-4. Schematic representation of laminar flow. F = Shearing force. V_1 to V_m = Velocity of the individual fluid layers $(1 - m)$ under the influence of the shearing force.

This law assumes laminar (Newtonian) flow of the liquid under the influence of shearing forces. Fibrous or rodlike macromolecules usually involve nonlaminar flow and therefore additional measurements are needed with special viscometers to obtain the viscosity coefficient of solutions containing them.

The Ostwald viscometer (Fig. 2-5) provides for the calculation of the relative viscosity (η_{rel}) by the following equation:

Equation 17: $$\eta_{rel} = \frac{\text{Outflow time of solution} \times \text{density of solution}}{\text{Outflow time of solvent} \times \text{density of solvent}}$$

The relative viscosity per se does not provide enough insight in regard to the behavior of the solute molecule. The specific viscosity as defined here is more useful in this respect.

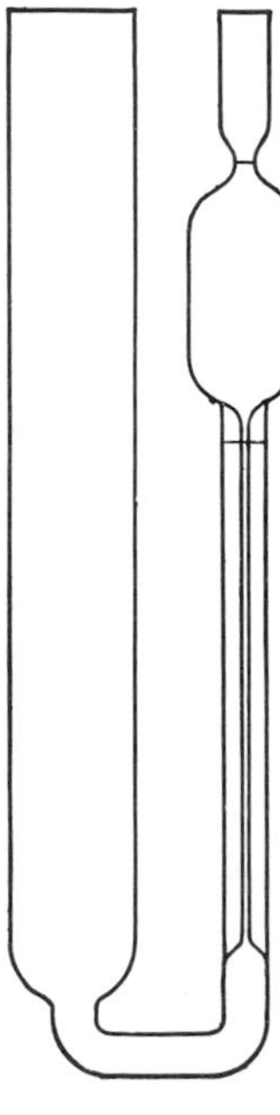

Figure 2-5. Ostwald viscometer. A definite volume of liquid is placed in the viscometer, and the level of the liquid is drawn above the top mark of the bulb by suction. The liquid is allowed to flow out freely, and the time required for the liquid level to drop from the upper mark to the lower mark is measured. (From Gortner, R. A., and Gortner, W. A.: Outlines of Biochemistry. 3rd edition. New York, John Wiley & Sons, 1953, p. 39.)

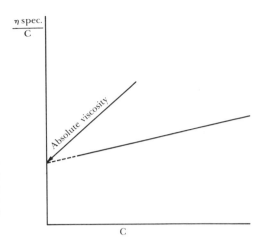

Figure 2-6. Concentration dependence of specific viscosity. The arrow points toward the intercept of the curve with the ordinate. The numerical value of this intercept is the absolute viscosity.

Equation 18: $$\eta_{\mathrm{spec}} = \eta_{\mathrm{rel}} - 1$$

η_{spec} = specific viscosity

By plotting specific viscosity divided by solute concentration against the solute concentration and extrapolating this value to zero solute concentration, the absolute viscosity of the solute is obtained; i.e., the viscosity coefficient becomes independent of solute-solute interactions (Fig. 2-6).

Equation 19: $$\lim_{c \to 0} \frac{\eta_{\mathrm{spec}}}{C} = [\eta]$$

$[\eta]$ = absolute viscosity

This parameter is very useful for calculations of the molecular weight and shape of macromolecules. If the molecular weight, the absolute viscosity, and the diffusion constant of a solute are known, the shape of this molecule may be estimated by the following equations:

Equation 20: $$\beta = \frac{D \cdot M^{1/3} \cdot [\eta]^{1/3} \eta_0 \cdot N}{RT}$$

or

Equation 21: $$\beta = \frac{S \cdot [\eta]^{1/3} \eta_0 \cdot N}{M^{2/3}(1 - \bar{v}\rho)}$$

$[\eta]$ = absolute viscosity of the solute
η_0 = relative viscosity of the solvent
N = Avogadro's number
β = a composite figure which reflects the ratio of the longitudinal (a) and horizontal (b) semiaxis of an ellipsoid of revolution. To make calculation possible the macromolecule is assumed to be an ellipsoid of revolution (see Fig. 2-7). If $a = b$ the shape is a perfect sphere.

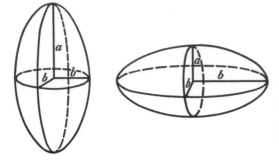

Figure 2-7. Ellipsoids of revolution. (a) Longitudinal semiaxis. (b) Horizontal semiaxis. (From Alexander, A. E., and Johnson, P.: Colloid Science. Oxford, Clarendon Press, 1949, p. 212.)

Light-Scattering Measurements and Macromolecular Structure

All materials capable of refracting light are able to scatter it. The light scattered by solute molecules diminishes the intensity of the incident light as it passes through the solution, a phenomenon quantitatively expressed by:

Equation 22:
$$I = I_0 e^{-\tau l}$$

I_0 = intensity of incident light
I = intensity of the emerging light
τ = turbidity
l = light path

The intensity of the light scattered by a particle may be expressed by the following equation:

Equation 23:
$$I_s = I_0 \frac{9\pi^2 \cdot V^2 \cdot n_1^2}{\lambda^4 X^2} \cdot \frac{n_2^2 - n_1^2}{n_2^2 + 2n_1^2} \cdot \sin^2\alpha$$

I_s = intensity of the scattered light
V = volume of the molecule which scattered the light
n_1 = refractive index of the solvent
n_2 = refractive index of the solution
α = the angle enclosed by the incident and scattered light beam
λ = the wavelength of the light used
X = the distance between the scattering particle and the light intensity measuring device

The scattering is caused by molecules becoming oscillating electric dipoles in response to the alternating electrical field of the incident light. The scattered light is therefore the product of the interaction of photons of light with the electrons of the scattering medium. The intensity of this scattered light, according to the equation, depends on the wavelength of the incident light and also on the volume of the scattering particle, the difference of refractive indices of solvent and solute, the distance between the scattering element and the detector, and the angle enclosed by the incident and scattered beams.

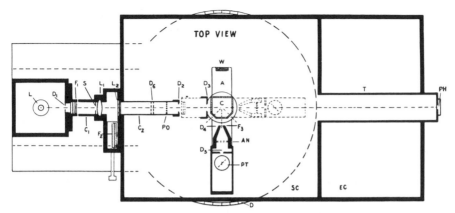

Figure 2-8. Top view of a Brice-Phoenix light-scattering photometer. The solution containing the macromolecule is placed in the scattering cell (C). The compartment which houses the scattering cell and the photomultiplier tube (PT) is light-tight and painted black. The light of a proper light source (L) is directed by lenses (L_1, L_2), diaphragms (D_1 to D_6), collimating tubes (C_1, C_2), and filters (F_1 to F_6) into the scattering cell. The angle between the incident light beam and the photomultiplier tube (PT) may be adjusted to the desired value and therefore the amount of light scattered by the solution may be measured in any desired angle. The figure shows the photomultiplier tube in a 90 degree angle from the incident light. Dotted lines show the photomultiplier tube in 0 degree angle position. T = Light trap tube. EC = Electric compartment. (From Phoenix Instruction Manual OM2000. Philadelphia, Phoenix Precision Instrument Co.)

Figure 2-8 is a sketch of a light-scattering instrument. Without respect to the angle at which scattering is measured, the total amount of light scattered by a solution is a function of the number of scattering molecules (Z) and the square of their volume (V^2).

Equation 24:
$$Z \cdot V^2 = \frac{c}{\rho} \cdot V$$

c = concentration of solute

Since by definition $c = V \cdot Z \cdot \rho$, or $\dfrac{c}{\rho} = V \cdot Z$

Equation 25:
$$c \cdot \frac{V}{\rho} = cM$$

In equation 23, I_o, λ, and X are instrumental constants; n_1 and n_2 are either known for a given macromolecular solution or are easily determined refractometrically. Furthermore at $\alpha = 90$ degrees the $\sin^2\alpha$ becomes 1; therefore in this special case the following expression is true.

Equation 26:
$$R_{90} = KcM \qquad \text{or} \qquad \frac{1}{M} = \frac{Kc}{R_{90}}$$

R_{90} = I of scattered light at 90 degrees
K = combined constants (X, λ, n_1, n_2, and π)

Equation 26 states the correlation between the intensity of the scattered light at 90 degrees and the molecular weight and concentration of the scattering particles. This correlation is valid only if the size of the scattering particle is less than one-twentieth of the wavelength of the light because above this size the particle does not behave as a point dipole. In order to exclude the effects of interactions between solute-solute and solute-solvent molecules, R_{90} has to be measured at various c values. A plot of $\dfrac{Kc}{R_{90}}$ against c yields a linelike curve with $\dfrac{1}{M}$ intercept on the y-axis (Fig. 2-9). It has been pointed out that this simple correlation is valid only if the size of the scattering particle is no larger than one-twentieth of the wavelength of the light used (e.g., if $\lambda = 4000$ Å, then maximum length of particle can be no greater than 200 Å). Until this length is reached the particles may be considered to be pointlike scattering elements. If the particle is much larger, the light scattered by one part of the molecule may be out of phase with the light scattered by another part of the same molecule. The interference of these scattered lights may lead to a diminished intensity compared to the ideal situation, i.e., interference-free scattering. Figure 2-10 shows that a hypothetical particle with two scattering elements may produce interference between the light beams scattered from the two independent elements. It is easy to see that the interference between the two scattered light beams may be destructive or constructive depending upon the relative position of the detector. Therefore in working with large scattering particles this effect has to be taken into consideration and the intensity of scattering must be determined at several different angles and with several solute concentrations at each angle.

Light-scattering measurements therefore offer a method for the determination of the molecular weight and also the shape of the protein molecules.

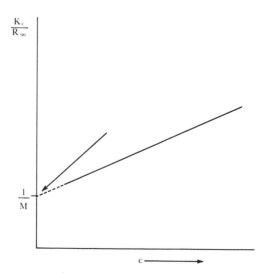

Figure 2-9. Determination of molecular weight of a spherical macromolecule by light scattering. The arrow points toward the intercept of the curve with the ordinate. The numerical value of this intercept equals the reciprocal molecular weight of the spherical macromolecule.

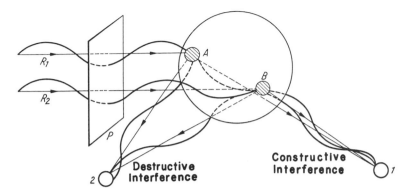

Figure 2-10. Light scattering of a large particle. *A* and *B* = Independent scattering elements. R_1 and R_2 = Light waves. (From Netter, H.: Theoretische Biochemie. Berlin, Springer-Verlag, 1959, p. 297.)

In order to obtain accurate information regarding the molecular weight, size, and shape of proteins, it is desirable that these parameters be calculated from data obtained by several independent methods. For example, the molecular weight of a protein (macromolecule) may be calculated by using five different approaches: (1) osmotic pressure measurement, (2) sedimentation equilibrium method, (3) light-scattering measurement, (4) determination of diffusion constant and sedimentation constant, and (5) determination of absolute viscosity and sedimentation constant.

The biological importance of these physical characteristics of the macromolecules is manifold. The molecular weight of an enzyme has to be known in order to calculate the efficiency of the enzyme molecule. To understand the organization and function of many subcellular structures (myofilament, membrane, chromosome, and so forth) the size and shape of its macromolecular components must be known.

Insofar as the elucidation of the correlation between structure and function is one of the most important goals of modern biology, the determination of the molecular parameters of biologically important macromolecules is exceedingly important.

Proteins as Electrolytes

Protein molecules possess a relatively large number of side chain groups (50 to 60 per 100,000 molecular weight) which may reversibly bind or release protons. The quantitative estimation of the proton-binding and proton-releasing ability of a protein yields useful information with respect to the structure and biological function of these macromolecules. With this approach the buffering capacity of a given protein can be assessed; moreover the shape of the titration curve gives an indication of the number and nature of the ionizing groups present and it also may reveal, along with

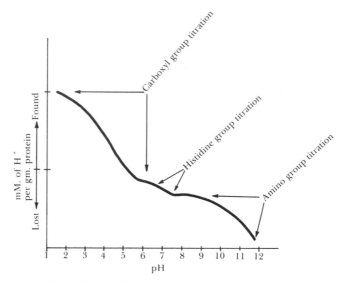

Figure 2-11.   Acid and base titration of a hypothetical protein.

amino acid analysis of the protein, the presence of some "masked groups."*
In selected cases it may be possible to correlate the appearance or disappearance of a charged group with an altered function of a protein, e.g., loss of enzyme activity or other biological property.

It may be seen from Table 1-1 in Chapter 1 that the pK of a specific ionizing group of an amino acid residue of a protein may differ from that found with the free amino acid. This relates to the fact that other amino acid residues as well as other components of the protein in the neighborhood of the ionizing group in question influence its dissociation.

In general, during titration the [H$^+$] is plotted on the x-axis and the total amount of hydrogen ions added to or removed from the protein is plotted on the y-axis. Figure 2-11 presents the titration curve of a hypothetical protein.

It may be seen from Figure 2-11 that the protein under consideration functions as a buffer between pH 3 and 6, with the carboxyl groups reversibly combining with protons. The protein also functions as a buffer between pH 6.3 and 8.0, with the imidazole groups of the histidine side chains reversibly combining with protons. Finally, at pH 9.0 and above the protein functions as a buffer because amino groups are reversibly combining with protons. It may be seen from Table 1-1 in Chapter 1 that proteins often contain ionizing side chains other than the ones seen in Figure 2-11. They also may be detected by titration. The interpretation of results at extreme alkaline and acid conditions is complex because proteins often denature in these pH ranges.

---

* There is discrepancy between the titration curve and the amino acid analysis; e.g., amino acid analysis may show ten tyrosine groups per mole of protein but the titration discloses the presence of eight. Two are "masked."

The interaction of proteins with small molecules is strongly influenced by the net charge on the protein as well as the charge on a specific amino acid residue. This phenomenon is of fundamental biological importance. For example, ion transport and enzyme action involve interactions between proteins and small molecules or ions.

REFERENCES

1. Bull, H. B.: An Introduction to Physical Biochemistry. Philadelphia, F. A. Davis Co., 1964.
2. Martin, R. B.: Introduction to Biophysical Chemistry. New York, McGraw-Hill Book Co., 1964.
3. Tanford, C.: Physical Chemistry of Macromolecules. New York, John Wiley & Sons, 1966.

3

THE
STRUCTURE
OF PROTEINS

THE FOUR ORGANIZATIONAL LEVELS OF PROTEIN STRUCTURE

The extremely complex nature of the three-dimensional protein structure and the importance of the integrity of this structure in the chemical and biological function of these macromolecules require a rather precise description of the structure itself. Linderstrom-Lang classified the organization of protein structure at four levels, primary, secondary, tertiary, and quaternary structure. It must be emphasized that such a classification does not imply that any level of organization exists in isolation; on the contrary, the levels are interdependent. Moreover the borderline between these various levels of structural organization is rather uncertain. The following definitions therefore must be viewed in a flexible way.

1. The primary structure of a protein relates to the number, nature, and sequence of the amino acid residues of the protein. For a protein consisting of several different polypeptide chains the primary structure involves the amino acid composition and sequence of the individual chains, e.g.,

α chain—alanine-valine-tyrosine . . . glycine
β chain—glutamic acid-tyrosine-glycine . . . alanine
γ chain—leucine-leucine-valine . . . aspartic acid

2. Secondary structure of proteins refers to the occurrence of any regular repeating structural arrangement or periodic foldings within the polypeptide chains. Helical structures and sheet formations belong to this category. The

secondary structure is thought to be mainly stabilized by hydrogen bonds between the carboxyl oxygen and amide nitrogen of the polypeptide chain.

3. The tertiary structure of proteins relates to the twisting or folding of the polypeptide chains other than those classified under secondary structure. The tertiary structure may be stabilized by a number of different linkages which do not repeat in any regular order. Such bonds include intrachain or interchain disulfide bridges (covalent bonds), hydrophobic or electrostatic interactions between the side chains of the polypeptide backbone, or even hydrogen bond formation at various points.

4. The quaternary structure defines the organization of interacting polypeptide chains, often termed the subunits of the protein. Therefore a protein "molecule" which possesses quaternary structure may be regarded as a polymer comprising low-molecular-weight proteins. The quaternary structure of a protein depends mainly on the surface interactions between the subunits.

DETERMINATION OF THE PRIMARY PROTEIN STRUCTURE

One of the greatest advances of modern biochemistry has been the development of methods suitable for the determination of the amino acid sequence of polypeptide chains. The first great landmark in this effort was the determination of amino acid sequence of the two polypeptide chains of insulin (M.W. 6000). For this work Sanger was awarded the Nobel Prize in 1958. Since then this methodology has been further developed and the amino acid sequence of an ever increasing number of proteins is being established.

The following is a brief résumé of the technique of determining the amino acid sequence of a protein.

1. The independent polypeptide subunits of proteins containing more than one polypeptide chain are separated and each is analyzed separately.

2. These individual polypeptide chains are converted to smaller polypeptide units.*

3. The amino acid sequence of the smaller polypeptide fragments is determined.

4. From a consideration of the overlaps of the smaller fragments it is possible to establish the unique amino acid sequence which must have existed in the parent polypeptide chain. This is usually done on the basis of "overlapping sequences."

5. If in steps 1 and 2 covalent bonds other than peptide bonds are broken, the nature and site of such linkages must be defined.

---

* The small polypeptide spectrum of a cleaved protein is often called the "fingerprint" of that protein.

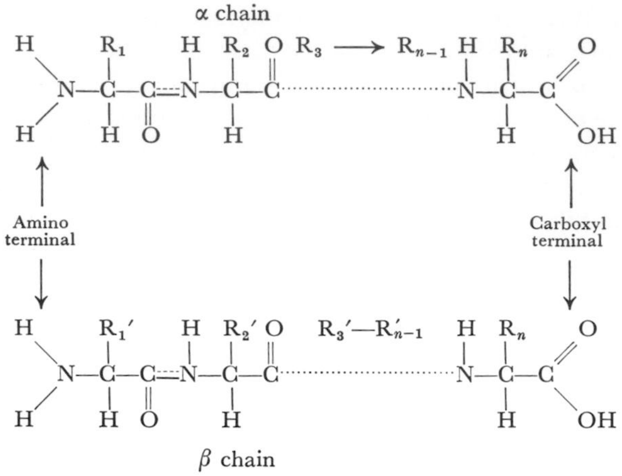

Figure 3-1.   Diagrammatic illustration of two separated polypeptide chains of a hypothetical protein. $R =$ Side chain.

Obviously the first step in the determination of amino acid sequence is to learn the number of independent polypeptide chains in the structure of the protein. Each polypeptide chain usually contains an amino acid with a free α-amino group at one end of the chain and an amino acid with a free α-carboxyl group at the other end. They are called, in the jargon of the field, the amino end group and carboxyl end group, respectively. Assuming that the terminal amino and carboxyl groups of the polypeptide chain are free, the determination of the number of amino terminal groups and carboxyl terminal groups per mole of protein will yield the number of independent chains. The example in Figure 3-1 has two polypeptide chains, each possessing different amino and carboxyl end terminal amino acid residues. The number of amino terminal amino acid residues per mole of protein may be determined by several methods. One such method involves interaction of the dinitrofluorobenzene with the free amino groups.

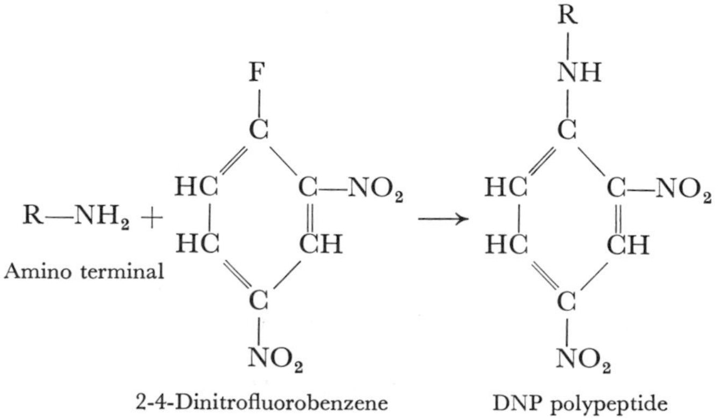

2-4-Dinitrofluorobenzene   DNP polypeptide

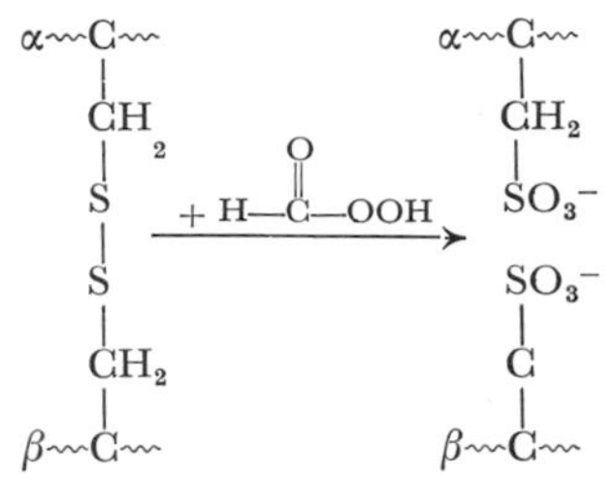

$$\alpha\sim\!C\!\sim \qquad\qquad\qquad \alpha\sim\!C\!\sim$$

$$\underset{2}{CH} \qquad\qquad\qquad CH_2$$

$$S \quad + H\!-\!\overset{O}{\underset{\|}{C}}\!-\!OOH \longrightarrow \quad SO_3{}^-$$

$$S \qquad\qquad\qquad SO_3{}^-$$

$$CH_2 \qquad\qquad\qquad C$$

$$\beta\sim\!C\!\sim \qquad\qquad\qquad \beta\sim\!C\!\sim$$

Figure 3-2. Disruption of a disulfide bridge by oxidation with performic acid.

The DNP tag is attached to the amino terminal group(s) of the polypeptide chain or chains and after hydrolysis the chemical nature of the DNP amino acid can be determined. Less satisfactory procedures are available for the carboxyl terminal group determination. The so-called hydrazinolytic method is frequently used. This method takes advantage of the fact that hydrazine interacts with the peptide bonds of the polypeptide chain, converting all amino acids with carboxyl groups in peptide linkage to acyl hydrazine derivatives. Obviously the terminal amino acids containing free carboxyl groups are not so converted; thus, by determining the chemical nature of the unmodified amino acids remaining after the treatment, the nature of the carboxyl terminal amino acids can be ascertained. After the number of polypeptide chains present is known, each kind of chain must be isolated free of all others. Since disulfide bridges often interlink the polypeptide chains, methods must be used which are able to break all interchain linkages without destroying peptide bonds. There are several methods for the cleavage of such bridges. One example is the oxidation of the —S— S— bridges with performic acid. This treatment transforms the cystine structure to cysteic acid (Fig. 3-2). After the individual chains have been isolated, the larger chains have to be cleaved to smaller fragments in order to determine the amino acid sequence of the individual polypeptide chains.

An isolated chain of the protein can be cleaved to smaller fragments by several methods. The smaller fragments obtained by the different methods differ but usually have overlapping parts (see Fig. 3-3). If the intermediary amino acid sequence of the individual smaller polypeptide chains is established, the areas of "overlapping sequences" are disclosed. If enough overlapping sequences are established, the unique amino acid sequence along the entire polypeptide chain is established with minimum ambiguity.

Both enzymatic and chemical procedures are available to cleave the long polypeptide chains with reasonable specificity into smaller polypeptides suitable for intermediary sequence determination. Trypsin cleaves peptide bonds in which the carboxyl groups of arginine or lysine are in peptide linkage with other amino acids (Fig. 3-3, I). If the polypeptide is treated by ethyl trifluorothioacetate before the trypsin action, only the arginine-associated peptide bonds are attacked by the enzyme; the peptide bonds involving lysine are no longer attacked because the ε-amino groups of the lysine are blocked by the reagent (Fig. 3-3, II).

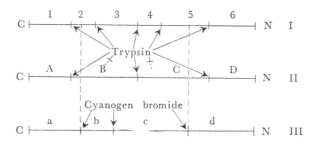

Figure 3-3. Fragmentation of a polypeptide chain by three different methods. I. Fragmentation of the polypeptide chain by trypsin. II. Fragmentation of the polypeptide chain by trypsin after trifluoroacetate treatment. III. Fragmentation of the polypeptide chain by cyanogen bromide.

Among the various chemical methods employed for this purpose, the most useful are those methods which cleave the polypeptide backbone more or less selectively at a given kind of side chain. One such method is the cyanogen bromide cleavage of the peptide bond at a methionine side chain (Fig. 3-4).

If the separation and isolation of smaller polypeptides is successful, the amino acid sequence of these compounds can be determined by several methods. A most useful one is the so-called Edman degradation procedure.

Enzymes are also used for sequence determinations. Two exopeptidases, carboxypeptidase and leucine aminopeptidase, yield this kind of information.

Knowledge of the primary structure of a protein molecule is extremely important. Small changes in the primary structure may lead to serious impairment of protein function. A well known example is the case of the abnormal hemoglobins, which will be discussed later in this chapter. Small differences in amino acid composition may change the biological activity of a protein hormone or even abolish it (see Chapter 8). A small chemical change in the primary structure may cause important alterations in the secondary and tertiary structure of the protein or may change the electrostatic charge at a given site of the macromolecule. These facts may often explain the profound biological consequences of a small chemical alteration of the primary protein structure.

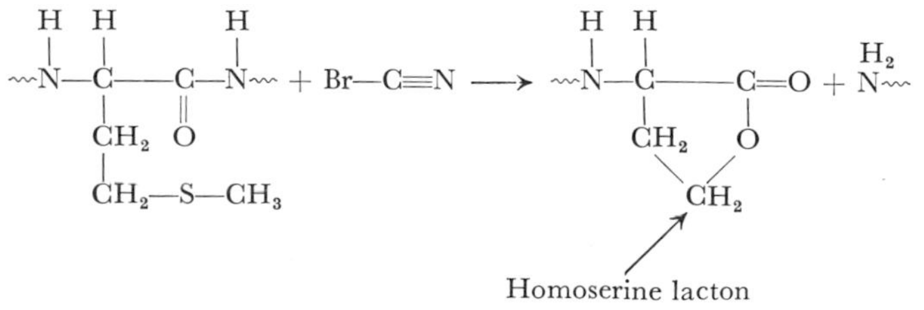

$$\begin{array}{c}\overset{H}{\underset{|}{N}}-\overset{H}{\underset{|}{C}}-\overset{H}{\underset{||}{C}}-\overset{H}{\underset{|}{N}} \end{array}$$

~N—C——C—N~ + Br—C≡N ⟶ ~N—C——C=O + N~
    CH₂ O   CH₂ O
    CH₂—S—CH₃   CH₂

Homoserine lacton

Figure 3-4. Chemical cleavage of a polypeptide chain by cyanogen bromide.

SECONDARY AND TERTIARY PROTEIN STRUCTURE

Bonds Involved in the Stabilization of Protein Structure

The stability of the three-dimensional protein structure depends on the nature and number of chemical bonds which link the various atoms together. For this reason the type and properties of some of the important bonds that participate in the protein structure will be briefly considered.

COVALENT BONDS. Covalent bonds contain bond energies greater than 35 Kcal. per mole. Therefore at least this much energy is needed to break the bond. The most important covalent linkages in proteins are the peptide bond and the disulfide bridges. The characteristics of the peptide bond have been studied in great detail. The most important qualities of this linkage have been established by x-ray crystallography and will be discussed in this connection later in this chapter. The peptide bond is the principal linkage between the amino acids within the polypeptide backbone and involves interaction of the α-carboxyl group of one amino acid with the α-amino group of another amino acid, with the elimination of a water molecule, as follows:

$$
R_1-C \begin{smallmatrix} COOH \\ \\ H \\ NH_2 \end{smallmatrix} + R_2-C \begin{smallmatrix} COOH \\ \\ H \\ NH_2 \end{smallmatrix} =
$$

$$
NH_2-\underset{H}{\overset{R_1}{C}}-\overset{O}{\overset{\|}{C}}-N-\underset{H}{\overset{R_2}{C}}-COOH + H_2O
$$

The disulfide bridge occurs between two cysteine residues located within the same polypeptide or in two different polypeptide chains of the protein. Figure 3-5 shows schematically the disulfide bridges linking the two polypeptide chains of the insulin molecule.

It may be seen that the α chain is linked to the β chain by two disulfide bridges and the configuration of the α chain is stabilized by a third intrachain disulfide bridge. The disulfide linkages function importantly in the stabilization of the tertiary structure of the proteins.

HYDROGEN BONDS. A hydrogen atom which is attached to an electronegative atom may interact with a second electronegative atom of the same or a different molecule. The lengths of the bonds so formed between the two atoms vary between 2.5 and 2.8 Å. Fluorine, oxygen, nitrogen, and sulfur, if located at a proper distance, may participate in hydrogen bond formation. The bond energy of this type of linkage is low, usually around 5 Kcal. per mole. In spite of this, hydrogen bond formation may be extremely important

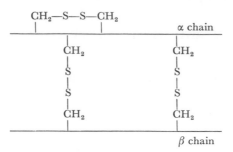

Figure 3-5. Schematic representation of the intrachain and interchain disulfide bridges in the insulin molecule.

in stabilizing the three-dimensional structure of proteins because of the large number of the potential hydrogen bond-forming groups. Indeed experimental evidence indicates the existence of intensive hydrogen bond formation between backbone groups and between side chain groups of proteins.

Some of the possibly important hydrogen bond types which may occur in the protein structure are:

$$R—OH \cdots O{=}C—R_1 \qquad R—NH \cdots O{=}C—R_1$$

Tyrosine   Carboxyl   Amino   Carboxyl

$$R—\overset{+}{N}H \cdots N—R_1$$

Histidine   Histidine$_1$

Hydrogen bond formation modifies the ionization of the groups involved in this linkage and also interferes with the interaction of the involved groups with small ions. Extensive hydrogen bond formation between protein molecules may enhance aggregation. Urea and guanidinium salts are often used to break hydrogen bonds in proteins. This is based on the strong hydrogen bond-forming ability of these molecules which may act as hydrogen donors and compete for acceptor groups on the protein molecule.

$$
\begin{array}{c}
\text{H} \\
\text{NH} \cdots \\
/ \\
\text{C}{=}\text{O} \\
\backslash \\
\text{NH} \cdots \\
\text{H}
\end{array}
\qquad
\begin{array}{c}
\text{H} \\
\text{NH} \cdots \\
/ \\
\text{C}{=}\text{NH} \cdots \\
\backslash \\
\text{NH} \\
\text{H} \cdots
\end{array}
$$

Urea   Guanidine

HYDROPHOBIC BONDS.   Figure 3-6 is intended to illustrate the principle of bond formation between apolar or hydrophobic protein groups. It may be seen that the presence of hydrophobic groups in the immediate vicinity of water molecules interferes with the hydrogen bond formation between

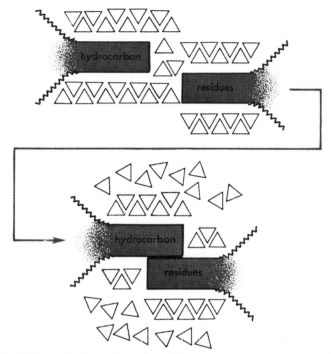

Figure 3-6. Hydrophobic bond formation. The triangles represent water molecules. On approach of the hydrocarbon residues to each other, some of the water molecules are released from an ordered arrangement close to hydrocarbon to a more disordered state in bulk solvent. (From Kopple, K. D.: Peptides and Amino Acids. New York, W. A. Benjamin, Inc., 1966, p. 92.)

water molecules and therefore modifies the orientation of adjacent water molecules. In other words, the structure of water in the immediate vicinity of these hydrophobic side chains is believed to be different ("icelike") from that in the rest of the liquid (liquid structure).

If a second hydrophobic side chain comes in close contact with the first apolar side chain (see arrow in Fig. 3-6), then it may be expected that the organized "icelike" water surrounding the apolar groups will be excluded along these hydrophobic sites of both the side chains in contact.

Since a certain number of water molecules are thereby transferred from a more organized to a more random state, such interaction is connected with an entropy increase and therefore proceeds spontaneously in the direction of hydrophobic bond formation. Leucine, isoleucine, valine, phenylalanine, proline, and methionine side chains may be involved in such bonding. Hydrophobic side chains of proteins orient preferentially inward. It is widely believed that hydrophobic bond formation is an important factor for the stability of protein structure.

Salt linkages and van der Waals bonding probably contribute less to the structural stability of proteins than the bonds discussed here.

Role of Optical Rotatory Dispersion in the Elucidation of Secondary Structure of Proteins

The determination of secondary and tertiary structure of proteins in solid state is readily accomplished by x-ray diffraction measurements. Fortunately there are other methods which provide information about the secondary and tertiary structure of proteins in solution. An important method is optical rotatory dispersion. That most amino acids contain an asymmetrical carbon atom and therefore rotate the plane of polarized light has already been discussed. Proteins are complex three-dimensional molecules built from these optically active amino acids. Accordingly, proteins also possess optical rotatory activity related to their optically active amino acid residues. The specific rotation of native proteins at λ-578 is between -30 and -80 degrees and that of the denatured proteins between -80 and -120 degrees. Polyglutamic acid exits as a random coil at neutral pH and as an α helical molecule at acidic pH. The transition of polyglutamic acid from the helical to the random coil configuration always involves an increased levorotation. Therefore, besides the composition and absolute number of the individual optically active residues (L-amino acids), the spatial arrangement of these residues within the molecule influences the optical activity of the macromolecule (Fig. 3-7). Obviously, analysis of optical rotation of proteins at different wavelengths (optical rotatory dispersion measurements) should provide information about the secondary structure of the polypeptide chain. The helical content of hemoglobin, myoglobin, and some other proteins was found to be nearly the same with optical rotatory dispersion and x-ray diffraction measurements.

A word of caution should be added. The rotatory dispersion measurement cannot disclose the amount of right-handed and left-handed α helical content separately, but only the difference between the right- and left-handed helical content. Therefore the method may yield an erroneously low value for the helical content.

Random coil

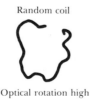

Optical rotation high

Helical coil

Optical rotation low   **Figure 3-7.**   Random and helical coils.

Study of Protein Hydrogen Bonds with Hydrogen-Deuterium Exchange

The role of hydrogen-deuterium exchange in the elucidation of hydrogen bonding within the protein molecule provides much information about the three-dimensional structure of proteins. The hydrogen atoms of protein molecules bound to nitrogen, oxygen, or sulfur exchange rapidly with the surrounding aqueous medium. However, those hydrogen atoms involved in hydrogen bond formation or those "buried" within the protein structure in such a way as not to be readily accessible to the solvent H_2O do not readily exchange H with the medium. Moreover hydrogen atoms linked to carbon exchange slowly. Studying the kinetics of the exchange of protein hydrogen atoms for the deuterium atoms of an aqueous medium enriched with deuterium oxide can provide useful information regarding the extent of hydrogen bonding within the protein structure.

Protein Conformation and Its Change During Denaturation

The secondary and tertiary structure of a protein is often called protein conformation, and hydrogen bonding plays an important role in it. Moreover biochemical functions of proteins relate to a specific conformation called the native structure. Therefore it is important to study conformational changes of proteins when investigating a phenomenon in molecular biology. Many of the physical-chemical tools discussed in this and earlier chapters can be used to provide information about protein conformation. In the case of enzymes the protein conformational change may involve the active center. In such cases kinetic analysis of the enzymatic activity may indirectly provide data on a possible conformational change. The conformation of a protein is determined both by its primary structure and by the environment in which it is located. Environmental changes may produce small and reversible conformational changes or major irreversible changes. Denaturation usually means that the protein conformation has been altered; such an alteration is often accompanied by loss of biological function of that protein (e.g., enzymatic activity). Denaturation is not necessarily an irreversible structural or functional change; e.g., reduction of the four —S—S— bridges to —SH groups causes a marked structural alteration of ribonuclease, with the complete loss of its enzymatic activity, but reoxidation of the —SH groups to —S—S— leads to the reappearance of full enzymatic activity. However, a similar treatment may cause irreversible alterations in some other enzyme. A protein is usually less soluble when it is denatured than in the native form, and the levorotation of the denatured form may be higher than that of the native conformation. The number of titratable groups often increases during denaturation. These changes are caused by the disappearance of helical structure of the native form and by cleavage of the hydrogen

bonds which stabilized the native structure. Because of this structural alteration the denatured protein is often more susceptible to the attack of proteolytic
enzymes than the native form.

UNIQUE ROLE OF X-RAY DIFFRACTION IN THE STUDY OF PROTEIN STRUCTURE

With proteins that crystallize and can be studied in the solid state,
x-ray diffraction analysis is the best method to elucidate the three-dimensional structure of macromolecules. The tertiary structure of the polypeptide
chain, which involves twisting and folding of the chains within the globular
or fibrous form of the molecule, is readily established by x-ray diffraction.

Since secondary and tertiary structures play most important biological
roles and are far more labile than the primary structure, it is in order at this
time to discuss some of the findings obtained with x-ray crystallography.
Before doing so, the techniques involved in x-ray diffraction measurements
will be mentioned.

Principles of X-Ray Diffraction Measurements

X-rays are generated when metals are bombarded with fast electrons,
which may expel inner-shell electrons from the bombarded metal. X-rays
are actually emitted by the metal when electrons from its outer shells fall
into the inner shell to fill the vacancies. The x-rays so produced are filtered
and collimated and may be used for diffraction studies (Fig. 3-8).

The wavelength of the x-rays is slightly less than 1 Å; the spacing of
atoms in molecules is of the same order of magnitude (e.g., $C—C = 1.53$ Å,
$C—N = 1.32$ Å). It is known that the electron density in a molecule is
greatest close to the nucleus of an atom and lowest between the atoms. X-rays
are diffracted by electrons and the diffraction pattern obtained from a molecule discloses the pattern of electron density within the molecule and therefore the spacing of atoms. Many macromolecules are composed partly or
entirely of periodically repeating units. From calculations based on three-
dimensional x-ray diffraction patterns the spatial relationship of the repeating
units and the spatial arrangement of the various atoms within these units
can be ascertained.

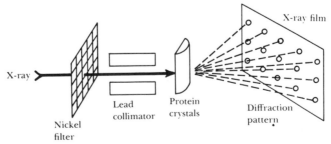

Figure 3-8.   Schematic illustration of some principal components which are necessary to
obtain x-ray diffraction patterns.

Fundamental Dimensions and Properties of the Peptide Bond

The polypeptide chain which forms the backbone of every protein molecule is composed of amino acids which are joined by peptide bonds. Some characteristics of the peptide links in a fully extended peptide chain are shown in Figure 3-9.

1. The fundamental dimensions such as bond length and angle of the peptide groups are independent of the side chains of the various amino acids.

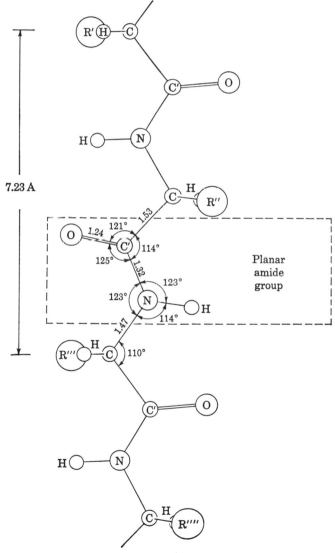

Figure 3-9. Dimensions of an extended polypeptide chain. (From Corey, R. B., and Pauling, L.: Proc. Roy. Soc. *B141*:10, 1953.)

2. The rotation around the C=N bond is restricted. This bond has 40 per cent of the character of a double bond. Therefore the amide group is planar.

3. The distance between two α carbons (C_α) is 3.3 Å.

4. The N—H groups are involved in interchain or intrachain hydrogen bond formation with the C=O residues if the C—N distance is 2.8 Å or less.

X-ray diffraction studies showed that long fibrous proteins are composed of ordered and disordered regions. The ordered regions resemble the crystalline structure in the sense that they consist of repeating units, whereas no such repeating units can be demonstrated in other regions of the polypeptide chain or chains.

THREE-DIMENSIONAL STRUCTURE OF FIBROUS PROTEINS BELONGING TO THE KERATIN GROUP

Some proteins obtained from hair (keratin), muscle (myosin), skin (epidermin), blood (fibrinogen), and other substances have certain common characteristics. They are quite extensible: after full extension on treatment with heat or alkali their length may be twice that in the contracted or α form.

Astbury studied the x-ray diffraction pattern of keratin and found in the α form a repeat period of 5.1 Å along the longitudinal axis of the chain which disappeared in the β form where a longitudinal repeat period of 3.38 Å was discovered.

The α Helix

From these x-ray diffraction results as well as from stereochemical and energetic considerations, Pauling and Corey formulated the concept of α helix for the contracted configuration of the keratin molecule.

In the α helix 3.6 residues are accommodated per 360 degrees. The pitch of the helix, i.e., the distance along the longitudinal axis of the helix per one turn, is 5.4 Å. Consequently the rise per residue is 5.4 per 3.6 Å. See Figures 3-10 and 3-12.

The stability of this structure is greatly enhanced by intrachain hydrogen bond formations which run parallel with the longitudinal axis of the helix. The hole within the corkscrew of the α helical chain is too small to accommodate any groups and therefore all the side chains are turned outward and are generally perpendicular to the longitudinal axis of the helix. The dimensions and structure of the β form (extended form) of keratin are similar to those represented in Figure 3-9.

The Compound Helix

As mentioned before, the longitudinal repeating unit distance in the α form of keratin was found to be 5.1 Å, which is considerably smaller than the 5.4 Å pitch of the α helical structure proposed by Pauling and Corey.

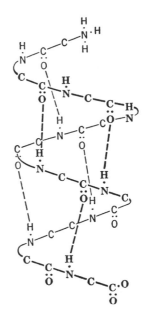

Figure 3-10. α Helix. The solid line shows the backbone of the polypeptide chain. The broken (dotted) lines depict the intrachain hydrogen bonds. (From Orgel, L. *in* Biophysical Science, A Study Program. Edited by J. L. Oncley. New York, John Wiley & Sons, 1959, p. 101.)

This difficulty can be overcome by assuming that the α helix is further bent into a superhelix which has a pitch approximately 12.5 times of that of the α helix. The longitudinal repeating units in such a coiled coil would be 5.1 Å apart. This superhelical structure (Fig. 3-11) would permit a firm packing of coiled interwound chains, which in fact were shown to be present in several fibrous protein molecules.

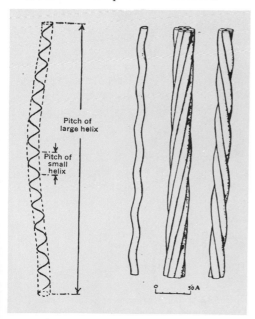

Pitch of large helix

Pitch of small helix

A   B

Figure 3-11. Compound helix. (*a*) Compound helix with pitch of the major helix equal to 12.5 times pitch of the minor helix. (*b*) *Left*, Compound α helix. *Center*, Seven-strand cable. *Right*, Three-strand rope. (From Pauling, L., and Corey, C.: Nature *171*:59, 1953.)

The Pleated-Sheet Structures

Pauling and Corey sought a theoretical model to explain the x-ray diffraction results as well as the chemical behavior of the proteins with the characteristic of the conformation of keratin. They deduced that adjacent polypeptide chains may be linked together by hydrogen bonds between the chains to yield a stable configuration. As Figure 3-12 shows, the hydrogen bonds in this case are formed laterally more or less perpendicular to the longitudinal axis to the adjacent chains. The side chains are perpendicular to the longitudinal chain axis, too, but they are also perpendicular to the plane of the book and so that they point toward and away from the reader. Pauling and Corey called these arrangements parallel-pleated sheets and antiparallel-pleated sheets, depending upon the alignment of matching points between the adjacent planes.

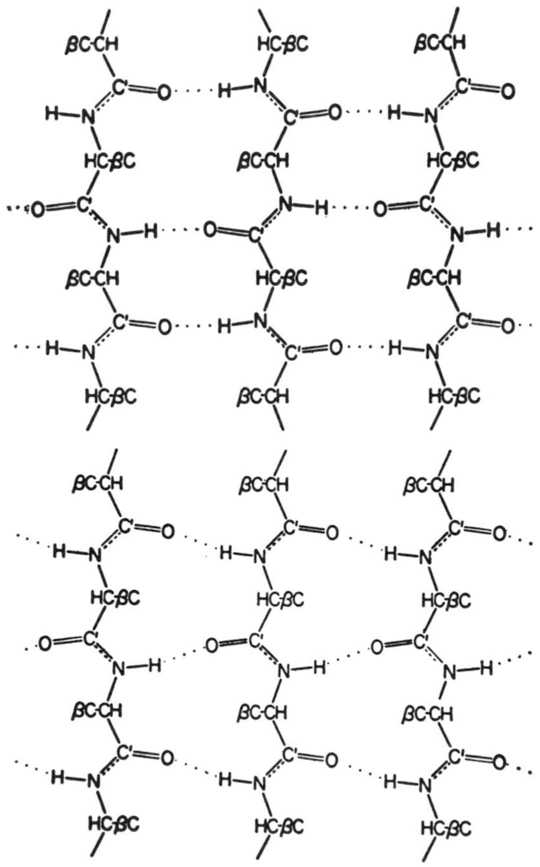

Figure 3-12.   Pleated-sheet structures. *Above,* Antiparallel arrangement. *Below,* Parallel arrangement. For explanation see text. (From Pauling, L., and Corey, C.: Proc. Nat. Acad. Sci. *37*:251, 1951.)

SOME STRUCTURAL CHARACTERISTICS OF PROTEINS OF THE COLLAGEN GROUP

Collagen is the principal protein of the tendon. This protein is much less extensible (about 11 per cent) than the proteins of the keratin group. Its x-ray crystallographic patterns and therefore its three-dimensional arrangement are different from those of the keratin group. The special structural characteristics of collagen may be due to the unique amino acid composition. Glycine makes up about 33 per cent of the protein and hydroxyproline is responsible for 20 per cent of the total amino acids in this group of proteins. Digestion of collagen with collagenase produces a large number of peptides. Amino acid sequence analysis of these peptides indicated that glycine-proline-hydroxyproline and glycine-proline-alanine are very common sequences and perhaps glycine is the third amino acid in each triplet. Bond rotations in the proline and hydroxyproline side chains are reduced and therefore the number of possible conformations for the peptide chain in which these imino acids participate is also limited. Accordingly, the large proline and oxyproline content confers great thermal stability in the collagens.

The Structure of Some Globular Proteins

Most globular proteins can be crystallized from solution. Their molecular weight is generally smaller than that of the fibrous proteins. They are therefore most suitable for precise x-ray diffraction analysis which in several instances has been successful in uncovering the details of the atomic arrangement of these molecules. Kendrew and Perutz were awarded the Nobel Prize for the determination of the three-dimensional structure of myoglobin and hemoglobin.

Myoglobin

Myoglobin is a small protein with a molecular weight of 17,000 and it contains one polypeptide chain. This protein functions intracellularly in the muscle tissue as an O_2 store. The O_2 is bound to the heme (ferrous protoporphyrin IX) molecule. The heme, which is attached to the myoglobin, consists of an iron atom forming coordinate linkages with the four pyrrole groups of the heme molecule (Fig. 3-13); the two remaining coordinates of the iron atom are connected with a histidine residue of the myoglobin molecule and with the molecular oxygen when the molecule is oxygenated. In this molecule 80 per cent of amino acid residues are arranged in a right-handed α helical form; the remaining 20 per cent give no sharp resolution with x-ray diffraction and most probably are present in the corners where the polypeptide chain twists and folds.

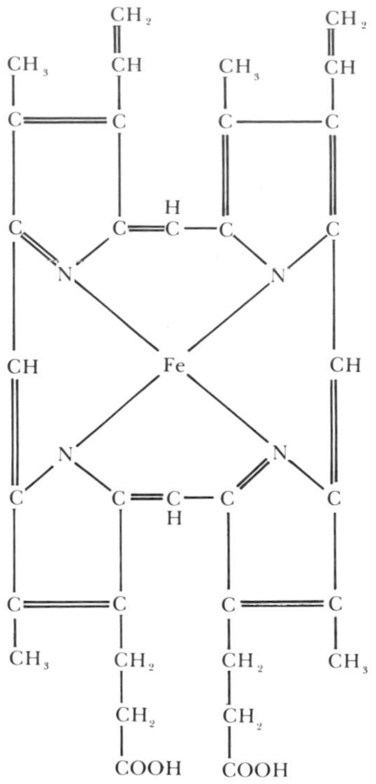

Figure 3-13. Structure of heme.

Hemoglobin

Hemoglobin has a molecular weight of 65,000. It is composed of four polypeptide chains, two α and two β chains, and four heme molecules are present. In spite of the fact that the primary structure of the hemoglobin polypeptide chains is known and the steric configuration of the chains is rather well understood, a precise explanation of its oxygen-binding capacity is still not available. Data have been collected to provide some understanding of the impaired oxygen binding in some pathological hemoglobins. In case of hemoglobin M disease, the hemoglobin is incompletely oxygenated, and cyanosis results. Chemical and spectroscopic studies indicate that the iron atom of the hemoglobin is in the ferric rather than the ferrous form and that it is directly bound to the pathological hemoglobin molecule. Studies on the primary structure of hemoglobin M indicated that histidine at position 58 of the normal hemoglobin α chain is replaced by a tyrosine, that histidine at position 63 of the normal β chain is replaced by a tyrosine, and that valine at position 67 of the normal β chain is replaced by a glutamic acid. It may be seen from Figures 3-14 and 3-15 that all the amino acid residues mentioned are close to the O_2-binding site. It is therefore possible that the phenolic group of the tyrosine or the γ-carboxyl of the glutamic acid forms direct links with the iron atoms and thus inhibits the activity of the enzyme methemoglobin reductase, thereby preventing the reduction of Fe^{+++} to Fe^{++} in the hemoglobin molecule.

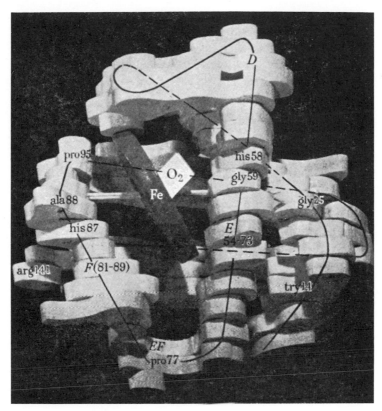

Figure 3-14. Back view of the α chain of hemoglobin. (From Cullis et al.: Proc. Roy. Soc. *A265:*15, 1962.)

Carboxypeptidase

In 1966 Lipscomb and co-workers established the three-dimensional structure of the bovine pancreatic carboxypeptidase (CPA). This is an enzyme which cleaves peptide and ester bonds at the carboxyl terminal residue. At this point the α-carboxyl of an L-amino acid is free, a requirement for CPA action, which is further enhanced if the carboxyl terminal amino acid has an aromatic side chain. The molecular weight of this enzyme is 34,600 and it contains 1 mole zinc per mole protein, an element essential for enzymatic activity. Figure 3-16 shows the structure of the enzymatic site. It may be seen that the zinc atom lies in a groove close to the surface of the molecule. Right at the zinc atom there is a pocket with a diameter of 8 to 10 Å leading to the interior of the molecule. Evidence indicates that the walls of this pocket are formed by hydrophobic amino acids. These components are believed to participate in the structure of the active site in the following way: The carboxyl oxygen of the peptide bond to be cleaved lies close to the zinc atom while the aromatic side chain of the carboxyl terminal amino acid is accommodated by the pocket.

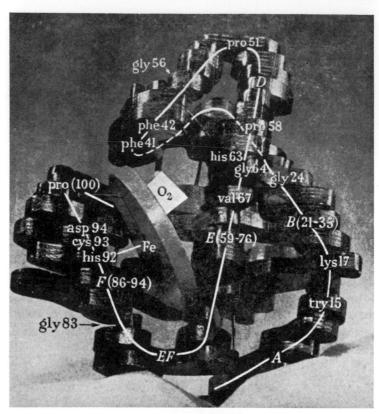

Figure 3-15. Back view of the β chain of hemoglobin. (From Cullis et al.: Proc. Roy. Soc. *A265:*15, 1962.)

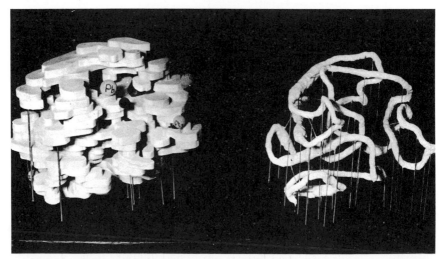

Figure 3-16. Models of the carboxypeptidase A molecule at 6 Å resolution, emphasizing the active site region around Zn^{++}. (From Lipscomb, W. N.: Structural Chemistry and Molecular Biology. San Francisco, W. H. Freeman and Co., 1968, p. 41.)

Obviously, precise knowledge of the three-dimensional structure will ultimately enable us to understand the molecular mechanism of fundamental life processes such as enzymatic action or muscle contraction.

REFERENCES

1. Harrington, W. F., Joseph, R., and Segal, D. M.: Physical and chemical studies on proteins and polypeptides. Ann. Rev. Biochem. *35:*599–650 (1966).
2. Neurath, H.: The Proteins. New York, Academic Press, Vol. 1, 1963; Vol. 2, 1964.
3. Perutz, M. F.: Proteins and Nucleic Acids. Structure and Function. New York, American Elsevier Publishing Co., 1962.
4. Rich, A., and Davidson, N.: Structural Chemistry and Molecular Biology. San Francisco, W. H. Freeman and Co., 1968.
5. Scheraga, H. A.: Protein Structure. New York, Academic Press, 1961.
6. Steiner, R. F.: The Chemical Foundation of Molecular Biology. Princeton, N.J., D. Van Nostrand Co., 1965.
7. Stryer, L.: Implications of X-ray crystallographic studies of protein structure. Ann. Rev. Biochem. *37:*25–50 (1968).
8. Timasheff, S., and Gorbunoff, M. J.: Conformation of proteins. Ann. Rev. Biochem. *36:*13–54 (1967).

4

METHODS OF ANALYSIS OF PROTEINS AND AMINO ACIDS

A large number of the chemical methods available for the determination of the protein concentration of a solution are not specific for the protein molecule but are based on interactions between a reagent and a constituent of the protein molecule. The most commonly used methods are described in this chapter.

COLOR REACTIONS OF PROTEINS AND AMINO ACIDS

Biuret Reaction

Dilute copper sulfate gives a blue-violet color reaction in the presence of sodium hydroxide with any compound which contains two —CO—NH— groups separated by a carbon or nitrogen atom.

Kjeldahl Determination of Protein Nitrogen

Assuming that the nitrogen content of all proteins is approximately 16 per cent by weight and that after digestion (boiling in concentrated H_2SO_4) all protein nitrogen is liberated and recovered, the Kjeldahl method of nitrogen determination provides a good quantitative assessment of protein, provided other nitrogen-containing compounds are absent.

Folin's Phenol plus Indole Analysis

The aromatic structures of tyrosine and tryptophan develop color when treated with the so-called Folin-Ciocalteu reagent in alkaline buffer. This reagent consists of sodium tungstate, sodium molybdate, and orthophosphoric acid. With this procedure the concentration of a given protein may be estimated on the basis of its tyrosine and tryptophan content. Lowry combined the advantages of this approach with those of the biuret procedure. His method is quite sensitive and reliable for the determination of protein concentration.

Ninhydrin Reaction

Free amino groups of amino acids or of proteins interact with ninhydrin, the nitrogen of the amino groups reacting with the reduced form of the ninhydrin reagent to produce a colored compound known as Ruheman purple. Step 1 is:

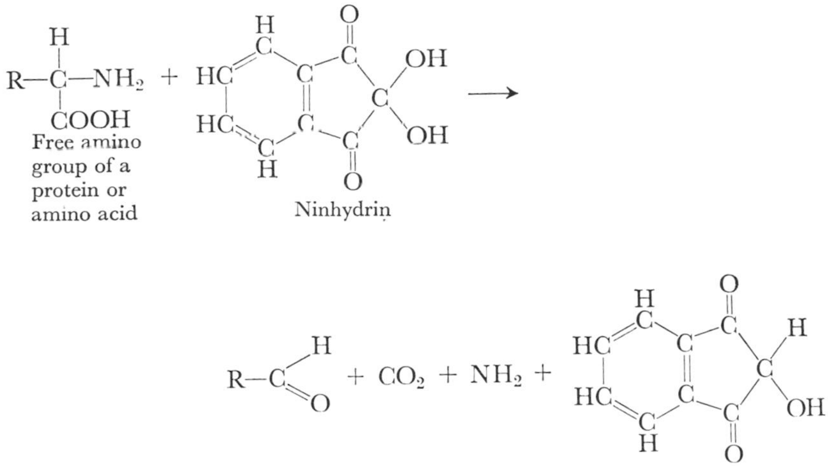

In the second step the colored compound is formed (see p. 56).

The depth of the color depends on the number of —NH_2 groups present and the nature of the amino acid residues. Therefore this procedure is not suitable to determine the amount of amino acid in a solution containing a mixture of these acids. If individual, separated amino acids are mixed with ninhydrin reagent, e.g., in automatic amino acid analysis, the observed color

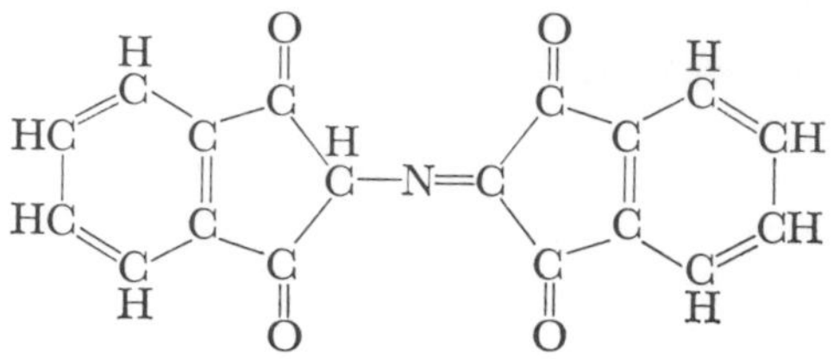

Ruheman purple

has to be corrected by an empirically derived factor characteristic for the kind of amino acid under consideration.

Ultraviolet Absorbance

The aromatic structures of tyrosine, tryptophan, and phenylalanine absorb light in the ultraviolet regions of the electromagnetic spectrum in the wavelength between 260 and 300 mμ. Care has to be taken that pH and temperature are constant and that there is no compound other than protein present which absorbs light in this ultraviolet region.

METHODS OFTEN USED FOR THE SEPARATION OF AMINO ACIDS, PEPTIDES, AND PROTEINS

Among the relatively large number of methods used for the separation and determination of these compounds, chromatographic techniques are currently the most popular. These methods not only are effective but usually are simple to perform; the equipment involved is readily available and not too expensive.

Chromatographic techniques require a fluid phase and a stationary phase. In this text, only those systems in which the fluid phase is liquid and the stationary phase is solid will be considered.

Column Chromatography

In these systems the stationary phase is packed in a column which is open at both ends and has an appropriate device on the bottom to keep the stationary phase from leaking out of the tubes. The liquid phase is permitted to flow through the column. The mixture undergoing chromatographic separation is usually applied to the top of the column. Then an eluent (a solvent called a developer) is perfused through the column. Since the components of the mixture usually have a different relative distribution between the fluid and the stationary phase, they migrate along the column at different rates. Therefore as these components migrate they are separated from each other; i.e., the components with higher affinities toward the solid phase will reach the lower part of the column much later than those with lower affinities for the solid phase (Fig. 4-1). This may be expressed by the following equation:

$$K = \frac{\text{Amount of substance attached to unit weight of stationary phase}}{\text{Amount of substance dissolved in unit volume of fluid phase}}$$

$$K = \text{equilibrium constant}$$

It is evident that the separation of substances with similar K values is likely to be less complete than separation of those with markedly different K values. Obviously the success of chromatographic separation depends upon the proper selection of phases so as to provide a favorable distribution of the K values among the individual components.

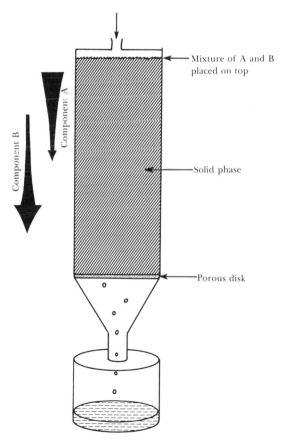

Figure 4-1. Schematic representation of a chromatographic column.

The interaction between the components of a solution and the stationary phase may be classified in the following way:

1. Direct interaction between the solid phase and the solute molecule (adsorption) (e.g., adsorption of nucleic acid to charcoal).

2. Partition of solute molecules between the moving liquid (fluid phase) and a second liquid phase held stationary by the solid. This type of chromatography is called partition chromatography.

3. Ion exchange processes involving the solute molecules and an insoluble acid or base (stationary phase). This ion exchange chromatography is widely used for the separation of the individual amino acids. It is also employed for the separation of peptides and nucleic acids. Since the amino acids (and also peptides and proteins) can exist in various ionic forms (depending upon the pH), their interaction with ion exchangers will vary with the pH. The varying interaction between individual amino acids and the ion exchangers of the stationary phases as the pH is continuously varied is the basis for the separation of the amino acids from their mixtures in automatic amino acid analyzers. Besides ionic structure, the nature of the amino acid side chains also plays a role in the chromatographic behavior of these molecules. Figure 4-2 shows schematically the principal components of automatic amino acid analyzers.

The amino acid mixture is fixed on top of the column and is eluted with a buffer mixture of varying pH. The fluid phase is mixed automatically with ninhydrin and the intensity of the color, which is dependent upon the nature of the amino acid and its concentration in the fluid phase at any given time, is measured by a photometer and plotted as a function of time. If the time needed for a given amino acid to be eluted under the standardized condition employed is known, the identity of every peak on the record is disclosed.

4. Molecular sieving: In this procedure the stationary phase contains porous particles. The solvent may freely penetrate into these pores but the extent of penetration of the solute depends upon the size and shape of the solute molecules. If the solute molecules are larger than the pore size of the stationary solid phase, they are excluded; if smaller, they are permitted to penetrate into the solid phase with the solvent. The result is that all those solute molecules which are "excluded" move fast, unopposed, through the column, whereas the other molecules, which are permitted to penetrate into the pores of the solid phase, are retarded and will leave the chromatographic column much later (Fig. 4-3).

This method is used extensively for the separation of proteins and polypeptides. Electrophoresis is discussed later in this chapter and ultracentrifugation is discussed elsewhere; both these methods are also used extensively to separate proteins, polypeptides, and peptides.

Paper Chromatography

This is a very simple and versatile form of chromatography. The stationary phase is filter paper and the fluid phase (or phases) is liquid.

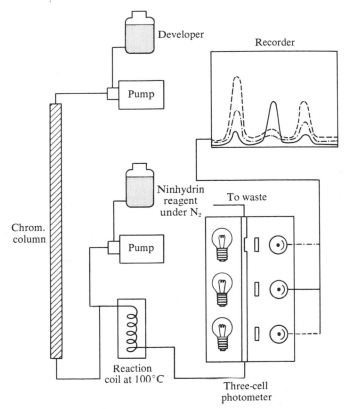

Figure 4-2. Principles of automatic amino acid analysis. The sample (amino acid mixture) is layered on top of the column. The liquid phase (developer) is pumped through the column into a reaction chamber. The ninhydrin reagent is mixed with the effluent liquid phase at 100°C. in the reaction chamber and pumped into the cuvettes of a photometer. The recorder plots the intensity of the color produced by the amino acid-ninhydrin interaction against time. The identity of the amino acids is established by the time needed for the appearance of individual peaks. The concentration of the individual amino acids is disclosed by the magnitude of their respective peaks. (From Schroeder, W. A.: The Primary Structure of Proteins. New York, Harper & Row, Publishers, 1968, p. 53.)

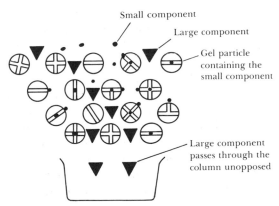

Figure 4-3. Schematic diagram of molecular sieving.

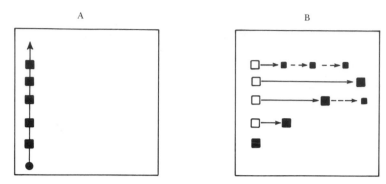

Figure 4-4. Two-dimensional paper chromatography. Fluid phases move in the direction of the arrows. *A*, First chromatography leads to partial separation of the components of the original mixture (●). Five spots are seen (■). *B*, Second chromatography produces further resolutions of the components of the original mixture. Eight spots are present (■).

The sample is placed on the paper close to one edge. After it dries (that is, after the sample is fixed), the liquid phase is permitted to penetrate through the filter paper, moving into the paper at the edge close to the sample. The various components of the sample are carried along by the liquid phase different distances from their original location as the fluid phase migrates. The components can then be detected on the filter paper, eluted from it, and further characterized. Paper chromatography may be used in one direction only or in two directions using two different liquid phases in two sequential steps. The chromatography with the second liquid phase is performed after the paper is rotated 90 degrees from the first phase. It may be seen from Figure 4-4 that during chromatography with the second solvent some of the individual spots obtained during the first chromatography are separated into several additional spots, indicating that they are composed of several distinct components not separated by migration of the first liquid phase.

Electrophoresis

One of the oldest known distinctive characteristics of proteins is their net charge. The mobility of a protein under the influence of electric field depends mainly on the surface charge of the macromolecule provided that other factors such as electric field force and temperature and viscosity of the medium are held constant. Electrophoresis involves the migration of proteins in electric fields and has been widely used to separate mixtures of proteins into component species, an approach which has been used successfully for analytical and preparative purposes. One method requires a careful layering of protein solution on top of the solvent and the monitoring of the moving protein boundary into the solvent. A proper optical system is provided to register the number, magnitude, and distance of migration of the various boundaries formed, and so the number, relative quantity, and mobility of the various components of a protein can be assessed. Needless to say, two different proteins having similar net charges and shape will migrate similarly in the electric field. Therefore more than one protein component may be present in any given boundary of an electrophoretic system.

Electrophoresis in supporting media such as paper, cellulose acetate, starch gel, and polyacrylamide gel is widely used. Figure 4-5 is a sketch of an electrophoresis apparatus using supporting media. The electrophoresis apparatus consists of a DC power supply and a chamber where separate places are provided for the buffer solution with electrodes and for a tray. The tray contains either the gel or the strips and is connected to the buffer chambers by some suitable device to ensure the flow of current.

The sample is applied in the form of a thin line on top of the strip or into the gel (without cutting the gel across and disturbing the continuity of the current flow). In the alkaline range most proteins carry net negative charge and they migrate toward the cathode. Electric current is permitted to flow through the system after a certain period of time and then the strip or gel is stained in proper medium so that the location of the various protein fractions on the supporting medium can be found. Since the combination of various proteins with the dyes used is a function of their respective concentration, the relative amounts of the various components may be evaluated.

Figure 4-6 shows the electrophoretic separation of the various serum protein fractions in a healthy person. This technique is an important and convenient tool in medical laboratories. The sera of patients suffering from various diseases exhibit different and often distinctive patterns of electrophoretic distribution of the various serum protein fractions. In a tumor-like disease of the bone marrow, myeloma multiplex, an abnormal protein fraction appears which migrates as a sharp band in the β or γ region or between these two regions (Fig. 4-7). In this case the altered electrophoretic pattern is characteristic for the disease and relates to the appearance of a new electrophoretically homogeneous protein produced by the tumor cells.

In cirrhosis of the liver, a chronic disease of the liver tissue connected with the increase of connective tissue and decrease of liver parenchyma,

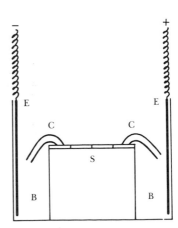

Figure 4-5. Schematic diagram of an electrophoresis apparatus using supporting media. E = Electrode. B = Buffer chamber. S = Supporting chamber. C = Connecting tube between the buffer chamber and the supporting medium.

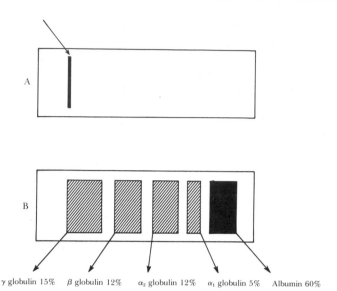

γ globulin 15%   β globulin 12% α_2 globulin 12% α_1 globulin 5%   Albumin 60%

Figure 4-6. Schematic paper electrophoresis pattern of a normal human serum. *A*, Layering the serum before the electrophoresis. *B*, The stained paper strip after the electrophoresis.

paper electrophoresis of serum proteins may show a fairly characteristic pattern. In this disease the γ-globulin fraction is markedly elevated and the appearance of a wide band suggests the presence of an electrophoretically inhomogeneous material in this band (Fig. 4-8).

Discoveries have been made during electrophoretic measurements which have provided insight into the mechanisms of the sickle cell anemia and some other anemias. The red blood cells of patients suffering from sickle cell disease undergo "sickling" in oxygen-poor media and hemolyze. Pauling and his collaborators found that the hemoglobin of normal adults migrates faster during electrophoresis than that of patients afflicted with sickle cell anemia (Fig. 4-9).

"Fingerprinting" experiments with the hemoglobin molecules (see Chapter 3) have shown that the hemoglobin S molecule (from sickle cell

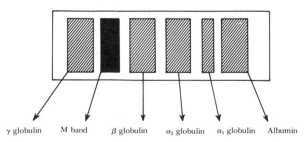

γ globulin   M band   β globulin α_2 globulin α_1 globulin   Albumin

Figure 4-7. Electrophoretic pattern of serum proteins of a patient with myeloma multiplex. M band = Myeloma protein.

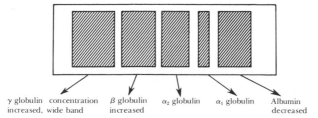

γ globulin concentration β globulin α₂ globulin α₁ globulin Albumin
increased, wide band increased decreased

γ globulin concentration increased, wide band   β globulin increased α_2 globulin α_1 globulin   Albumin decreased

Figure 4-8. Electrophoretic pattern of serum proteins in cirrhosis of the liver.

patients) differs from hemoglobin A (from normal people) by one amino acid. A glutamic acid molecule is replaced by a valine in this disease. The glutamic acid residue causes the normal hemoglobin to have one more negative charge than hemoglobin S at the neutral and alkaline pH range and thus explains the slightly higher rate of electrophoretic migration of the normal molecule. Paper electrophoretic measurements of hemoglobins are valuable aids for the diagnosis of several abnormal hemoglobinemias.

Paper electrophoretic measurements are also used for the assay of lipoproteins and carbohydrate-containing proteins.

Electrophoretic separation of serum proteins on various supporting media may be followed by tests for enzyme activity of the various separated protein fractions. It has been found that occasionally several distinct electrophoretic fractions may have similar enzymatic activity. Such findings have been related to the tissue of origin of these enzymes of similar catalytic function but different electrophoretic mobility. The molecular, genetic, and clinical (see also Chapter 6) significance of these enzymes, often called isoenzymes, is currently under intensive study.

PURIFICATION OF PROTEINS

Purification of a specific protein can be difficult because proteins are easily changed from their native state, i.e., denatured. The three-dimensional

A

B

Figure 4-9. Paper electrophoresis of hemoglobins. *A*, Normal hemoglobin. *B*, Hemoglobin of a sickle cell anemia patient.

structure of the purified protein should be the same as in the native state; if not, the protein is denatured. If the protein has a biological function the preservation of this biological activity during purification provides strong evidence that the native state is being maintained.

It is not possible in the space available to consider comprehensively the field of protein purification. Rather, a brief outline of general approaches will be presented.

In the case of intracellular proteins, the protein can be liberated by rupturing the cell by grinding, ultrasonic treatment, or freeze-drying. The cell homogenate can then be fractionated by differential centrifugation and the cell fraction containing the desired protein harvested. If the desired protein is contained in an isolated organoid, e.g., the mitochondria, the organoid must be disintegrated so as to free the protein. In such cases chemical agents such as detergents are often required to solubilize the protein. There are numerous methods which can be used to isolate specific soluble protein from other proteins, e.g., chromatographic techniques, ultracentrifugation, and electrophoresis. One of the oldest effective techniques is to precipitate certain proteins and to keep others in solution by varying the salt concentration or pH or both. Organic solvents and elevated temperature have also been used and are successful when the desired protein is not denatured by the procedure, but many of the other proteins are rendered insoluble by denaturation. The purity of the isolated protein can be assessed by chromatographic, electrophoretic, or ultracentrifugal behavior. Crystallization is a good but not foolproof index of purity.

REFERENCES

1. Alexander, P., and Block, R. J. (eds.): A Laboratory Manual of Analytical Methods of Protein Chemistry. Vols. 1, 2, and 3. New York, Pergamon Press, 1960.
2. Alexander, P., and Lundgren, H. P. (eds.): A Laboratory Manual of Analytical Methods of Protein Chemistry. Vol. 4. New York, Pergamon Press, 1966.
3. Leggett Bailey, J.: Techniques in Protein Chemistry. 2nd edition. New York, American Elsevier Publishing Co., 1967.
4. Schroeder, W. A.: The Primary Structure of Proteins. New York, Harper & Row Publishers, 1968.

5

CLASSIFICATION OF PROTEINS

Proteins which yield only amino acids on hydrolysis are called simple proteins. Proteins which yield amino acids and other chemical substances on hydrolysis are called conjugated proteins.

Simple Proteins

A time-honored and widely used classification of proteins is based on their solubility properties. Although the distinction between various subclasses on this basis is not sharp, it is useful for organizing the proteins into various families.

ALBUMINS. These proteins are soluble in water, salt solutions, and dilute acid or alkali. Albumins are widely distributed in living matter, and more than 50 per cent ammonium sulfate saturation is necessary to cause them to precipitate out of solution.

GLOBULINS. Globulins are insoluble (euglobulins) or poorly soluble (pseudoglobulins) in pure water at their isoelectric point. Small amounts of salt, dilute acid, or alkali usually dissolve these proteins. This solubilizing effect of salt (salting in) is usually attributed to the effect of salt on the charged protein groups. By partially neutralizing the surface charges of the protein, the salt diminishes the attraction between protein molecules and thus the protein is solubilized. In greater salt concentrations, in the case of globulins usually 0.2 to 0.5, saturation with ammonium sulfate precipitates these proteins. This effect (salting out) is explained on the basis of competition between the salt molecules and protein for the solvent.

PROLAMINES. The proteins belonging in this subclass are not soluble in water or in salt solutions but are readily dissolved in 50 to 80 per cent aqueous ethanol solutions. Most prolamines are isolated from plants.

GLUTELINS.   These proteins are insoluble in neutral aqueous solutions but dissolve readily in dilute mineral acids or alkali.   Glutelins are also obtained from plants and are only poorly characterized.

PROTAMINES.   Protamines are water-soluble and very small proteins (M.W. about 6000).   They are extremely rich in arginine (70 to 80 percent) and contain no sulfur.   The predominantly basic character of these proteins enables them to form salt with nucleic acids or other acid proteins.

HISTONES.   These proteins, like protamines, dissolve in water and contain much basic amino acid (lysine and arginine).   Their molecular weight is somewhat higher than that of the protamines.   They are found in combination with nucleic acids in biological systems.

ALBUMINOIDS OR SCLEROPROTEINS.   The proteins belonging to this subclass are found in various organisms mainly as structural proteins. Representative members of this group are the keratins from skin and collagen from tendons.   The solubilization of these proteins is a difficult and not completely resolved problem.   They are insoluble in most generally used solvents.   Some of their structural characteristics have been discussed in connection with protein structure.

Conjugated Proteins

LIPOPROTEINS.   Lipoproteins constitute a large, diverse, and biologically and medically important class of the conjugated proteins.   In the serum they migrate with the α, β, and γ globulins.   The protein to lipid ratio, as well as the nature of the lipid component, is different in the various subclasses of the lipoproteins.   In this text only the involvement of lipoproteins in the coagulation of blood will be discussed.   A more complete description of the chemistry and biological importance of lipoproteins can be found in *Physiological Chemistry of Lipids in Mammals* of this series.

MUCOPROTEINS.   These proteins are conjugated with carbohydrate. The variable carbohydrate moiety contains glucosamine combined with some of the following compounds: galactose, rhamnose, mannose, glucuronic acid, gluconic acid, uronic acid, sulfuric acid, and phosphoric acid.   Mucoproteins are found in tissues of the mammals and other species.   Their chemistry and biological importance will be discussed in *Physiological Chemistry of Carbohydrates in Mammals* of this series.

NUCLEOPROTEINS.   The prosthetic group in these conjugated proteins is nucleic acid.   They will be considered in Chapter 7.

CHROMOPROTEINS.   Chromoproteins constitute an ill defined subclass of the conjugated proteins in which protein is associated with a pigment. Hemoglobin, ceruloplasmin, hemocyanin, and so forth belong to this class.

REFERENCE

1. Greenberg, D. M. (ed.): Amino Acids and Proteins. Springfield, Ill., Charles C Thomas, 1951.

CHAPTER

6

ENZYMES

GENERAL DESCRIPTION

A catalyst is a substance which increases the speed of a thermodynamically possible chemical reaction. The catalyst is not a product of the reaction, nor does it change the equilibrium between reactants and products, although it shortens the time required for equilibrium to occur.

Enzymes are specific proteins produced by living organisms which function as catalysts for a specific reaction or class of reactions. A striking quality of enzymes is their specificity regarding the reactions they are able to catalyze. Emil Fischer drew an analogy between enzyme-substrate specificity and the perfect fit which must exist between a lock and its key. This specificity is a property of the complex protein structure of the enzyme. The amino acid residues or prosthetic groups of the protein surface involved in the interaction with the substrate* are called the active site. The chemical nature, number, and steric configuration of these groups determine the specificity of an enzyme and also its ability to function as a catalyst. Chemical alterations of functional groups at the active site may change the specificity of the enzyme. More often chemical modifications at the active site cause loss of enzymatic activity. At the present time a certain number of chemical modifying agents are available which are fairly selective for a given amino acid residue and thus make it possible to correlate the chemical modification of an amino acid residue with the change in enzymatic activity.

MECHANISM OF ENZYME ACTION

Enzymes catalyze only chemical reactions that are thermodynamically possible. The enzyme itself does not contribute energy to the reaction. The

---

* Substrate is a molecule which after interaction with a specific enzyme undergoes a chemical alteration.

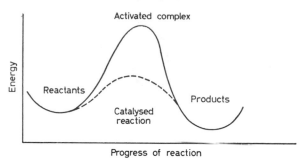

Figure 6-1. Energy profile of a reaction and the effect of a catalyst on the energy of the activated complex. (From Gutfreund, H.: An Introduction to the Study of Enzymes. Oxford, Blackwell Scientific Publications, 1965, p. 23.)

way in which an enzyme increases the rate of reaction may be considered in the following manner:

In Figure 6-1 the energy content of the reactants is higher than that of the products; therefore this reaction may and will occur spontaneously. Since, however, the reactants have to be converted to an intermediate compound before the products will form, the energy content of the intermediary compound may be so high as to almost completely stop the reaction from occurring, as is the case in Figure 6-1. In other words, at any given time only a very minute fraction of the reactant molecules possess a high enough energy content to be able to form the intermediate compound and thus only this very small fraction of the reactant molecules will yield the products at that time. The energy required for the reactants to form the intermediate is called the energy of activation. Let us assume that the reactants are small molecules and that the intermediary compound production requires the formation of a bond between the reactants. If the reaction is permitted to proceed in the absence of a catalyst or enzyme, the reactants may collide "unsuccessfully" a million times until one successful collison leads to bond formation between them, with production of the intermediate and its decomposition to the products. The reason for this is that the proper atoms of the reactant molecules must come close enough and assume a sterically favorable position in order that a new bond may be formed between them. When the reaction proceeds at random and the reactant molecules rotate around their own bonds, only a small fraction of the reactant molecules will reach that "sterically favorable position" necessary for the intermediate formation in a given time.

The enzyme possesses a specific site on its surface comprised of two or three amino acid side chains in a specific position. This site is able to line up the reactants in a relatively fixed position favorable for the bond formation between the reactants. Formation of the new bond between the reactants and the cleavage of another bond which permits the final product to emerge may proceed simultaneously. Part of the energy required for new bond formation between reactants may come from the energy of another bond being cleaved within the same intermediate molecule. This mechanism is probably one way that an enzyme is able to reduce the activation energy required and thus enhance the rate of the reaction. The long-range electrostatic forces of the charged groups must permit the adsorption of the substrate, which may be further facilitated by the contribution of London forces and hydrophobic interactions between enzyme and substrate. The formation of a covalent bond between enzyme and substrate has also been shown.

CLASSES OF ENZYMES

HYDROLASES. These enzymes catalyze the hydrolytic cleavage of certain bonds.

TRANSFERASES. The function of these enzymes is the transfer of a chemical group from a donor to an acceptor molecule.

OXIDASES, REDUCTASES. This class of enzymes transfers electrons or hydrogen atoms from a donor to an acceptor molecule.

ISOMERASES. These enzymes catalyze intramolecular rearrangements of the substrate without adding atoms to or subtracting them from that molecule.

CONDENSING ENZYMES. These enzymes catalyze the attachment of one molecule to another by the formation of covalent bonds.

VELOCITY OF ENZYME-CATALYZED REACTIONS

Many factors influence the rate of enzymatic reaction, the most important of which will be briefly discussed here.

$$v = k[E]$$

v = velocity of reaction

k = velocity constant, velocity of enzymatic reaction if $[E]$ is unity

$[E]$ = concentration of enzyme

The reaction velocity is proportional to the enzyme concentration, provided the substrate concentration is not a limiting factor (curve A, Fig. 6-2). In the case of "substrate exhaustion" the velocity does not increase in the presence of larger enzyme concentrations (curve B, Fig. 6-2). However, in practice it is often found that the role of an enzymatic reaction is not proportional to the enzyme concentration; this may be caused by the presence of inhibitors in the enzyme preparation which cause the velocity of the enzymatic reaction to increase at a rate less than expected from the increase in enzyme concentration (curve D, Fig. 6-2). Another cause of lack of proportionality between enzyme concentration and reaction velocity is the presence of small amounts of a toxic impurity in the assay medium (e.g., a heavy metal or detergent). Since only a small amount of the enzyme added is poisoned,

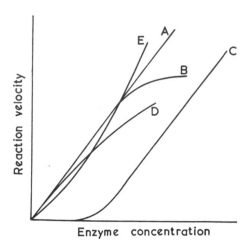

Figure 6-2. Effect of increasing enzyme concentration on amount of substrate transformed in unit time. A = Normal response. B = Effect of substrate exhaustion. C = Effect of toxic impurity. D = Effect of inhibitor in enzyme preparation. E = Effect of activator in enzyme preparation. (From Wilkinson, J. H.: Introduction to Diagnostic Enzymology. Baltimore, The Williams & Wilkins Co., 1962, p. 24.)

a smaller percentage of enzyme molecules will be affected as enzyme concentration increases, resulting in curve C of Figure 6-2. If as enzyme concentration is increased the increased reaction velocity is disproportionately great, an activator is probably present in the enzyme (see curve E of Fig. 6-2).

An equally important factor in determining the velocity of enzymatic reactions is the substrate concentration. If one plots the velocity of a reaction as a function of the substrate concentration, a curve similar to curve A, Figure 6-3, may be obtained.

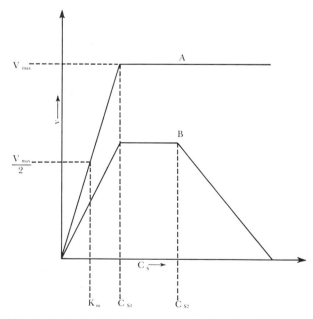

Figure 6-3. Plot of velocity against substrate concentration. C_S = Substrate concentration. v = Velocity of reaction. V_{max} = Maximal velocity. K_m = Michaelis constant.

It may be seen from curve A that between 0 and C_{S1} substrate concentration the reaction velocity increases rapidly. The further increase of substrate concentration above C_{S1}, however, is not connected with any increase in the reaction velocity. In other words, at C_{S1} substrate concentrations the reaction velocity becomes maximal (V_{max}) and remains constant if substrate concentration is increased further. The situation in the case of the enzymatic reaction depicted by curve B is somewhat different. Here too the reaction velocity initially increases as substrate concentration increases. At C_{S1} substrate concentration the maximal velocity is reached, but the maximal velocity declines as substrate concentration is increased further (C_{S2}). This phenomenon is called substrate inhibition.

The explanation for the velocity rise between 0 and C_{S1} relates to the fact that in this concentration range the active sites of the enzyme molecules are not fully saturated with substrate and therefore not all the enzyme molecules present can participate fully in the reaction. At C_{S1} full saturation of the enzymes with substrate has been achieved and the system works at maximal velocity. The decrease in velocity at high substrate concentration (C_{S2} in case of curve B) may be explained on the basis of an inhibition of the product desorption from the active site of the enzyme caused by the larger amount of substrate. In some cases large substrate concentrations may chemically alter the active site, resulting in inhibition.

The interaction between enzyme and substrate may be described the following way:

Equation 1: $$C_E + C_S \underset{k_2}{\overset{k_1}{\rightleftharpoons}} C_{ES} \xrightarrow{k_3} C_E + C_P$$

k_1, k_2, and k_3 = velocity constants of the corresponding reactions
C_E = total enzyme concentration
C_S = total substrate concentration
C_{ES} = total concentration of enzyme-substrate complex
C_P = total product concentration

This equation implies that the enzyme-substrate interaction consists of at least three reactions. The first reaction leads to the formation of an enzyme-substrate complex and the second and third reactions involve the decomposition of the enzyme-substrate complex (Fig. 6-4).

The rate of formation and decomposition of the enzyme-substrate complex is more precisely described in the form of differential equations.

Equation 2: Rate of C_{ES} formation $+ \dfrac{dC_{ES}}{dt} = k_1(C_E - C_{ES})C_S = V_1$

Equation 3: Rate of decomposition $- \dfrac{dC_{ES}}{dt} = k_2 C_{ES} + k_3 C_{ES} = V_2 + V_3$

$C_E - C_{ES}$ = concentration of free enzyme (total enzyme concentration minus the concentration of enzyme-substrate complex)

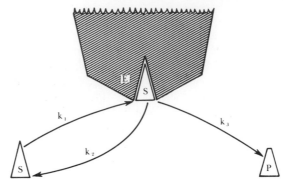

Figure 6-4. Simplified diagram representing the formation and decomposition of the enzyme-substrate complex.   S = Substrate.   E = Enzyme.   P = Product. k_1 to k_3 = Velocity constants.

V_1 = velocity of the reaction leading to enzyme-substrate complex
      formation

V_2 = velocity of the decomposition of enzyme-substrate to yield free
      enzyme and free substrate

V_3 = velocity of decomposition of enzyme-substrate complex to yield
      free enzyme and product

If these three reactions are in a state of dynamic equilibrium the rate of formation of the enzyme-substrate complex is equal to the rate of its decomposition.   In this steady state, equations 2 and 3 may be rewritten the following way:

since

$$+\frac{dC_{ES}}{dt} = -\frac{dC_{ES}}{dt} \qquad \text{(by definition of dynamic equilibrium)}$$

Therefore

$$k_1(C_E - C_{ES})C_S = k_2 C_{ES} + k_3 C_{ES}$$

and

Equation 4:
$$\frac{(C_E - C_{ES})C_S}{C_{ES}} = \frac{k_2 + k_3}{k_1} = K_m$$

K_m = Michaelis constant

The Michaelis constant (K_m) combines three velocity constants.   It is the ratio of the summed velocity constants of the decomposition reactions and the velocity constant of the formation reaction of the enzyme-substrate complex.   In other words, in this context the Michaelis constant is the dynamic equilibrium constant of the aforedescribed three reactions.   Therefore in the steady state enzyme-substrate complex is continuously being formed and decomposed partly back to substrate but partly to product (Fig. 6-4).

In this steady state $k_1 = k_2 + k_3$ and therefore $k_3 < k_1 > k_2$. Both k_1 and k_3 are involved in product formation (Figure 6-4 and equation 1). Since k_3 is always smaller than k_1, the decomposition of the enzyme-substrate complex to product will be the rate-limiting step for product formation. With this consideration as a basis, the velocity of the substrate conversion to product may be simply written as

Equation 5: $$V = k_3 C_{ES}$$

Equation 4 may be rearranged the following way:

Equation 6: $$K_m = \frac{(C_E - C_{ES})C_S}{C_{ES}} = \frac{C_E C_S - C_{ES} C_S}{C_{ES}}$$

Equation 7: $$K_m = \frac{C_E C_S}{C_{ES}} - C_S$$

After solving equation 7 for C_{ES} we obtain:

Equation 8: $$C_{ES} = \frac{C_E C_S}{K_m + C_S}$$

Substituting equation 5 into equation 8 we obtain:

Equation 9: $$V = \frac{k_3 C_E C_S}{K_m + C_S}$$

According to this equation the conversion of substrate to product by an enzyme depends on the enzyme and substrate concentrations, the Michaelis constant of the reaction, and the velocity constant of the breakdown of the enzyme-substrate complex to product.

As was mentioned before, at high levels of substrate concentration all available enzyme is saturated with substrate. Under these conditions the total amount of available enzyme is present as enzyme-substrate complex. Therefore if $C_E = C_{ES}$, then C_{ES} reaches its maximal value. According to equation 5, $V = k_3 C_{ES}$; if C_{ES} is at its maximal value, V has to be maximal as well.* Under these conditions equation 5 may be written as:

Equation 10: $$V_{max} = k_3 C_{ES} = k_3 C_E$$

$$V_{max} = \text{maximal velocity}$$

Combining now equations 9 and 10:

Equation 11: $$V = \frac{V_{max} C_S}{K_m + C_S}$$

---

* Usually C_S is $\gg$ than C_E in vitro and therefore one may assume that the numerical value of C_S decreases insignificantly while all C_E is transformed to C_{ES}.

Rearranging equation 11 we obtain:

Equation 12: $\dfrac{V_{max}}{V} C_S = K_m + C_S$ Michaelis-Menten equation

If $\dfrac{V_{max}}{V} = \frac{1}{2}$, then

Equation 13: $\dfrac{V_{max}}{\frac{1}{2} V_{max}} C_S = K_m + C_S$

$$C_S = K_m$$

We have now reached an experimentally useful definition of the Michaelis constant, K_m; specifically, it is numerically equal to that substrate concentration where the velocity of substrate-product conversion is half the maximal velocity.

The general definition for the Michaelis constant is, therefore, the following: K_m is a dynamic equilibrium constant of at least three reactions and its numerical value is equal to that substrate concentration at which substrate-product conversion has reached its half-maximal rate (see Fig. 6-3).

The Michaelis constant under specified conditions may have a meaning somewhat different from the one just described. If the decomposition of the enzyme-substrate complex to product is much smaller than its dissociation to substrate and free enzyme, then $k_3 \ll k_2$. Therefore equation 4 may be rewritten as:

Equation 14: $\dfrac{(C_E - C_{ES})C_S}{C_{ES}} = \dfrac{k_2 + k_3}{k_1} \equiv \dfrac{k_2}{k_1} \equiv K_m \equiv K_S$

K_S = substrate dissociation constant

Under these conditions the Michaelis constant is the true thermodynamic dissociation constant of the enzyme-substrate complex and its reciprocal is the true measure of affinity of the substrate to the enzyme ($K_m = K_S$). The Michaelis constant is one of the most important parameters of enzymatic reactions. It may not be easy to decide experimentally in a specific case whether or not the Michaelis constant is a true thermodynamic dissociation constant. In spite of the fact that the Michaelis constant has a rather complex kinetic meaning, it often reflects the affinity of a given substrate to the active site of the enzyme.

Between the adsorption of substrate and desorption of product several intermediate enzyme-substrate and enzyme-product complexes are possible.

Equation 15: $C_E + C_S \underset{k_2}{\overset{k_1}{\rightleftharpoons}} C_{ES} \underset{k_4}{\overset{k_3}{\rightleftharpoons}} C_{EP_1} \underset{k_6}{\overset{k_5}{\rightleftharpoons}} C_{EP} \underset{k_8}{\overset{k_7}{\rightleftharpoons}} C_E + C_P$

k_1 to k_8 = kinetic constants
C_{EP_1} = concentration of an enzyme intermediate complex constants
C_{EP} = concentration of an enzyme product
C_P = concentration of product

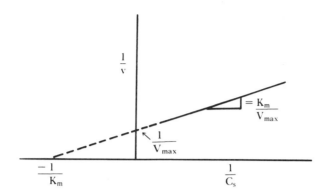

Figure 6-5. Plot of $\frac{1}{V}$ against $\frac{1}{C_s}$ according to Lineweaver-Burk. V = Velocity. C_s = Substrate concentration. V_{max} = Maximal velocity. K_m = Michaelis constant. (From Harrow, B., and Mazur, A.: Textbook of Biochemistry. 9th edition. Philadelphia, W. B. Saunders Co., 1966, p. 102.)

Therefore the conversion of substrate to product on the enzyme surface may be a rather complex process and its rigorous evaluation may not be possible by any assumption based on a simple mass law.

A simple mathematical rearrangement of the Michaelis-Menten equation proved to be very useful for the practical evaluation of its components. Equation 12 may be rewritten in the following form:

$$\frac{V_{max}}{V} = \frac{C_S + K_m}{C_S}$$

Dividing both sides by V_{max} we obtain:

Equation 16: $$\frac{1}{V} = \frac{K_m}{V_{max}}\left(\frac{1}{C_S}\right) + \frac{1}{V_{max}}$$

If $\frac{1}{V}$ is plotted against $\frac{1}{C_S}$ a straight line is obtained. The intercept on the y-axis is equal to $\frac{1}{V_{max}}$ and the slope of the line corresponds to $\frac{K_m}{V_{max}}$. This graphical evaluation is called the Lineweaver-Burk plot (Fig. 6-5).

EFFECT OF pH ON ENZYMATIC REACTIONS

Effective enzymatic catalysis requires a high degree of specificity between the active site and the substrate. Such specificity is provided by the steric configuration of the active site of the enzyme protein. By changing the pH of the reaction mixture the number and nature of the ionizing groups

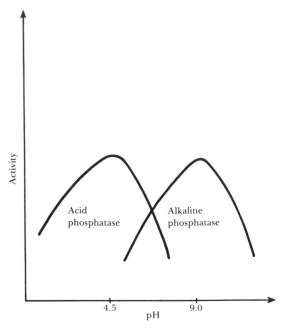

Figure 6-6.   pH versus activity plot of acid and alkaline phosphatases.

may be altered. This change alone may influence enzymatic reactions. For example, if the pH is increased from 6 to 8 histidine and sulfydryl groups may become ionized. If such groups are part of the active site, the probability for enzyme-substrate complex formation may be distinctively altered by their ionization. Indeed such effects of pH play a theoretically important part in the consideration of the nature of the active site.

The influence of pH change on the enzymatic reactions may be the consequence of factors other than the change in ionization of amino acid residues at the active site. For example, the electrostatic attraction or repulsion between enzyme and substrate may be altered as the consequence of ionization of certain side chains on the protein molecule as a whole. Moreover the structural stability of a protein at a given temperature is affected by the pH. A protein which is relatively stable for 30 minutes at pH 7.0 at 37°C. may be denatured in 10 minutes at pH 9.0. In a given case it may be difficult to decide unequivocally in connection with a pH-produced change in enzymatic rate to what extent a change in the ionization of the active site was responsible for the altered enzymatic activity.

The pH dependence of the enzymatic reaction is of great practical importance. If the proper pH is not ensured, the activity of enzyme may remain negligible. Different enzymes capable of catalyzing similar chemical reactions may have different pH optima. For example, among the phosphatases which catalyze the dephosphorylation of various organic phosphates, one enzyme is optimally active at an acid pH and the other at an alkaline pH (Fig. 6-6). In this particular case the difference in the pH optima of these enzymes has biological and diagnostic significance. In general, kinetic characterization of an enzyme is preferentially at its pH optimum. Needless to say, if the quantity of an enzyme is to be determined on the basis of its enzymatic activity, the pH of the reaction mixture is of utmost importance.

EFFECT OF TEMPERATURE ON ENZYMATIC REACTIONS

The rate of chemical reactions increases with increasing temperature. Therefore it is to be expected that the rate of enzymatic reactions will be greater as the temperature increases. However, enzymes are proteins and the structural integrity of proteins is temperature related, denaturation occurring at higher temperatures. These two factors tend to influence the enzymatic rate in opposite ways. Although there are appreciable variations from enzyme to enzyme, as a crude approximation it may be assumed that between $+17°$ and $37°C$. the rates of most enzymatic reactions increase with increasing temperature. Within this range, a $10°C$. increase in temperature usually doubles the reaction rate. Above $45°C$. the rate of reaction usually declines markedly (Fig. 6-7).

In selected cases the temperature dependence of the maximal velocity and Michaelis constants permits the calculation of the enthalpy change of a particular enzymatic reaction.

It is obvious that the temperature control of enzymatic reaction is fundamentally important since even small temperature changes may alter drastically the rate of reaction. Therefore in every case in which the amount

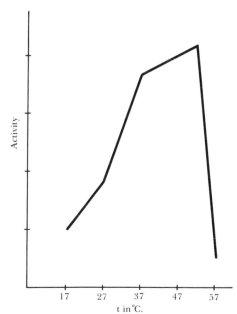

Figure 6-7. Activity against temperature plot of a hypothetical enzyme.

of enzyme or available enzymatic activity is to be determined the temperature of the reaction media has to be held meticulously constant.

COENZYMES AND ACTIVATORS

A large number of enzymes require a factor, or factors, in addition to the enzyme and substrate molecules for the catalytic reaction to occur. These additional factors may be organic molecules of various complexity or simple metal ions. Like the enzyme molecule, these cofactors can be recovered after the substrate-product conversion. Since they are essential for the enzymatic reaction and are not consumed during the process, they are called coenzymes.

Some coenzymes are firmly attached to the active site; others form loose complexes with it. Coenzymes which are firmly bound to the protein molecule are often called prosthetic groups.

The coenzymes have various mechanisms of action. In general, it may be said that the velocity of most enzymatic reactions which require a coenzyme is proportional to the coenzyme concentration within a certain coenzyme concentration range, and beyond that point a steady reaction velocity is achieved (Fig. 6-8). It may be assumed that at this concentration the enzyme becomes saturated with coenzyme. A similar reaction has been mentioned between substrate concentration and enzymatic reaction velocity.

As an example of the functioning of coenzymes, the deamination and transamination reactions which play an important role in the amino acid metabolism may be mentioned. Pyridoxal phosphate is a coenzyme required for the function of many enzymes participating in this process. In this case the aldehyde group of the protein-bound coenzyme forms a Schiff base with

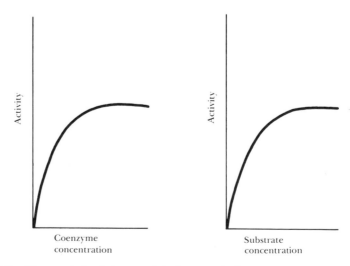

Figure 6-8. The correlation between substrate concentration, coenzyme concentration, and activity in the case of a hypothetical enzyme.

the amino donor (B). After rearrangement and hydrolysis (C, D), the aminated coenzyme, pyridoxamine, interacts with the amino acceptor molecule (E). This way the aldehyde group of the pyridoxal is regenerated and the coenzyme is ready to start the cycle again.

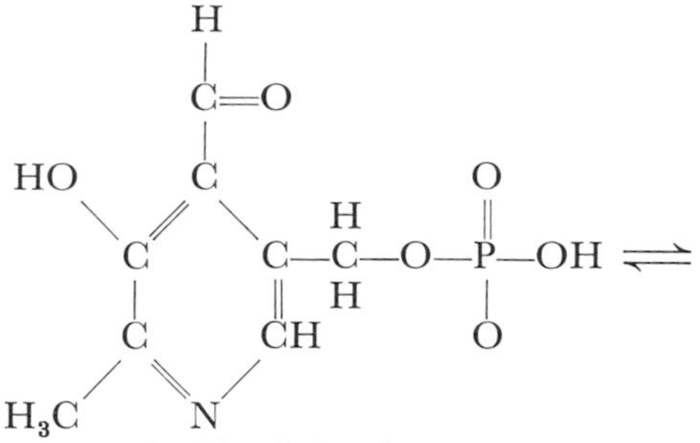

Pyridoxal phosphate

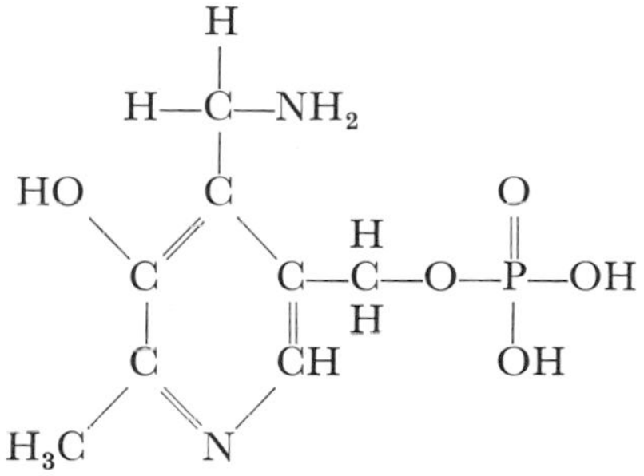

Pyridoxamine phosphate

The overall reaction:

A

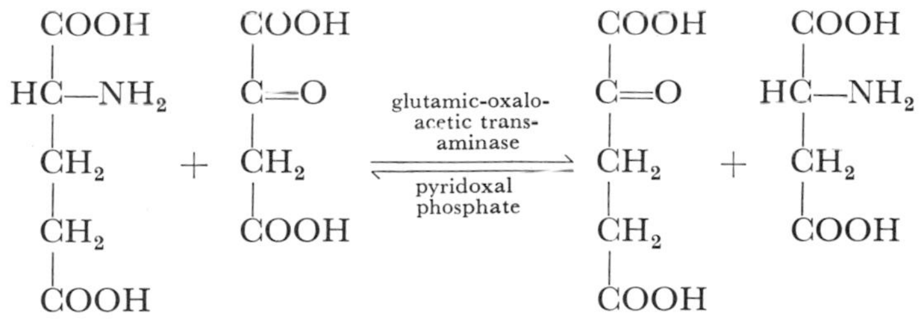

| Glutamic acid | Oxaloacetic acid | | α-Ketoglutaric acid | Aspartic acid |

First step:

B

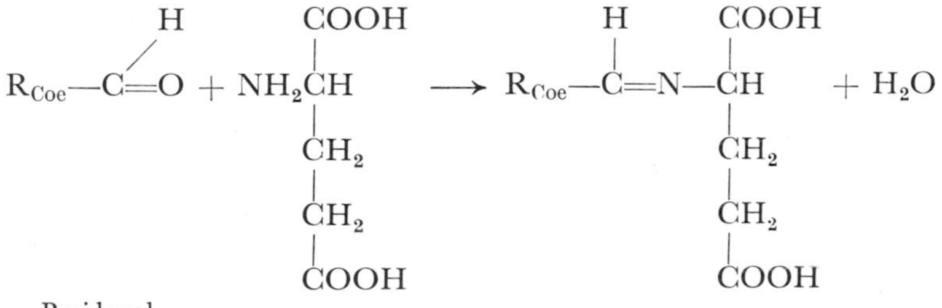

Pyridoxal phosphate Glutamic acid Schiff base

Second step:

C

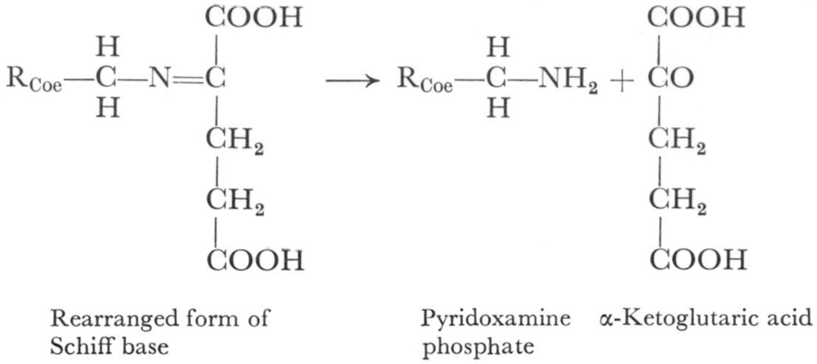

$$R_{Coe}-\overset{\overset{\displaystyle H}{|}}{\underset{\displaystyle H}{C}}-N=\overset{\overset{\displaystyle COOH}{|}}{\underset{\underset{\displaystyle COOH}{|}}{\underset{\displaystyle CH_2}{\overset{|}{C}}}} \longrightarrow R_{Coe}-\overset{\overset{\displaystyle H}{|}}{\underset{\displaystyle H}{C}}-NH_2 + \overset{\overset{\displaystyle COOH}{|}}{\underset{\underset{\displaystyle COOH}{|}}{\underset{\displaystyle CH_2}{\overset{|}{C}O}}}$$

      Rearranged form of   Pyridoxamine   α-Ketoglutaric acid
      Schiff base   phosphate

Third step:

D

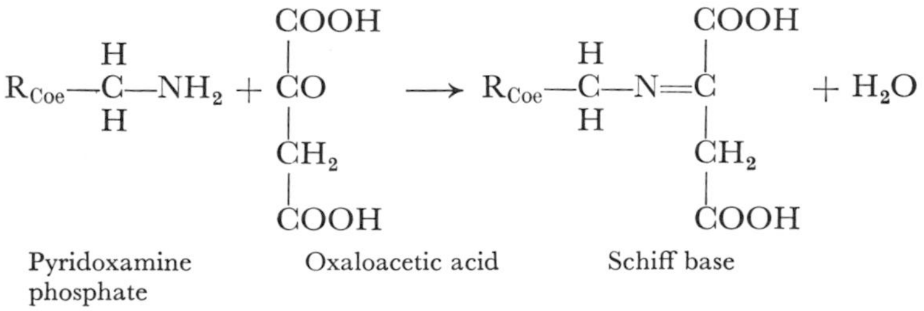

$$R_{Coe}-\overset{\overset{\displaystyle H}{|}}{\underset{\displaystyle H}{C}}-NH_2 + \overset{\overset{\displaystyle COOH}{|}}{\underset{\underset{\displaystyle COOH}{|}}{\underset{\displaystyle CH_2}{\overset{|}{C}O}}} \longrightarrow R_{Coe}-\overset{\overset{\displaystyle H}{|}}{\underset{\displaystyle H}{C}}-N=\overset{\overset{\displaystyle COOH}{|}}{\underset{\underset{\displaystyle COOH}{|}}{\underset{\displaystyle CH_2}{\overset{|}{C}}}} + H_2O$$

      Pyridoxamine   Oxaloacetic acid   Schiff base
      phosphate

Fourth step:

E

$$R_{Coe}-\overset{\overset{\displaystyle H}{|}}{C}=N-\overset{\overset{\displaystyle COOH}{|}}{\underset{\underset{\displaystyle COOH}{|}}{\underset{\displaystyle CH_2}{\overset{|}{CH}}}} + H_2O \longrightarrow R_{Coe}-\overset{\overset{\displaystyle O}{\|}}{C}-H + H\overset{\overset{\displaystyle COOH}{|}}{\underset{\underset{\displaystyle COOH}{|}}{\underset{\displaystyle CH_2}{\overset{|}{C}NH_2}}}$$

      Rearranged form of   Pyridoxal phosphate   Aspartic acid
      Schiff base

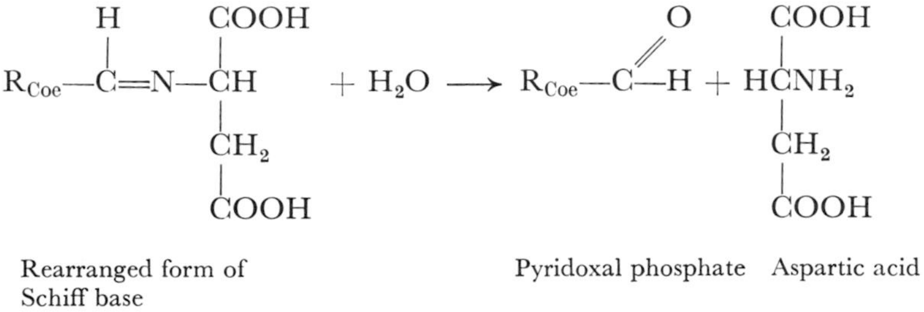

    Another example is the coenzyme or family of coenzymes which plays an important role in the synthesis of the purine and pyrimidine ring and some amino acids—the folic acid group. This group is not known to be attached firmly to any enzyme but rather functions in a manner similar to a carrier which shuttles back and forth between donor and acceptor. A very important characteristic of the folic acid, or rather the tetrahydrofolate, carrier coenzymes is that the group which is attached to this carrier may undergo chemical changes while being transported. This means that the acceptor molecule may not receive the same group that was attached to the tetrahydrofolate carrier by the donor. Folic acid must be reduced to tetra-hydrofolic acid to be able to serve as a coenzyme.

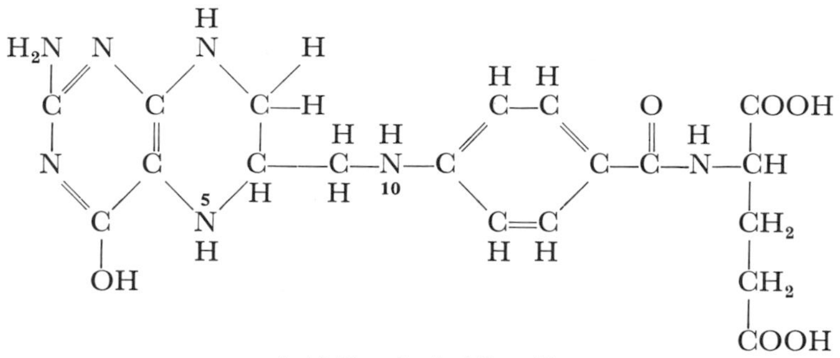

5–10-Tetrahydrofolic acid

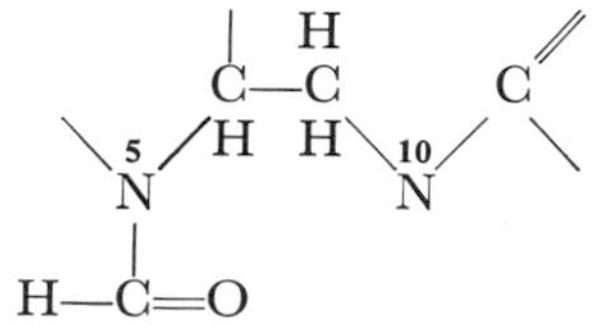

5′-Formyl tetrahydrofolic acid 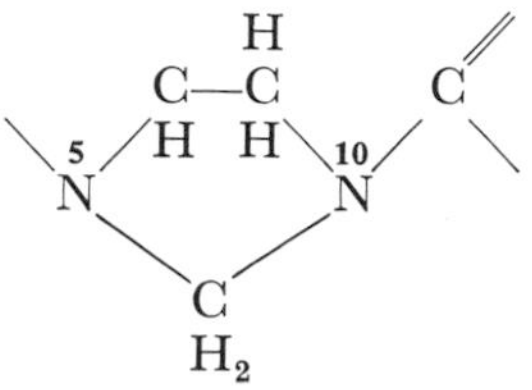   5′,10′-Methylene tetrahydrofolic acid

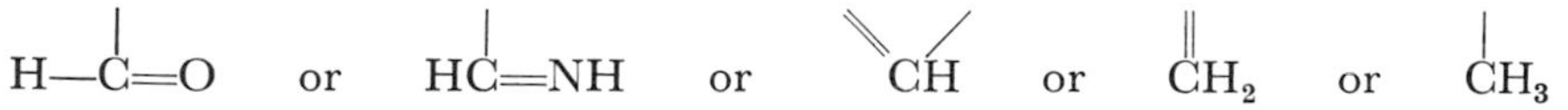

H—C=O   or   HC=NH   or   CH   or   CH$_2$ or   CH$_3$

Formyl   Formimino   Methenyl   Methylene   Methyl

Depending upon the donor and enzymes present, any one of the formyl, formimino, methenyl, methylene, or methyl groups may be attached to the nitrogen or nitrogens in position 5 or 10 or both. In the presence of another set of enzymes and other cofactors these attached groups may be interconverted and delivered to the proper one-carbon acceptor in the presence of a third set of enzymes.

Obviously coenzymes may function as a firmly attached group to the active site of a given enzyme or as a link between enzyme systems without being strictly attached to any given enzyme.

In either case the coenzyme becomes a part of the active center of the enzyme since it plays a major role in the conversion of the substrate to product. It has been mentioned that for maximal activity the enzyme has to be saturated with coenzyme. Usually there is a one to one ratio between enzyme and coenzyme. The actual concentration of the coenzyme at saturation depends on the affinity constant of the enzyme-coenzyme complex.

INHIBITION OF ENZYMATIC ACTIVITY

A compound which decreases the rate of an enzymatic reaction is called enzyme inhibitor. In vivo the organism utilizes inhibitors to regulate the velocity of the enzyme-catalyzed reactions of the life processes. In vitro inhibitors are used to study the active site of an enzyme as well as the mechanism of the enzymatic reaction. Enzyme inhibitors are also an important part of modern pharmacology. The rational experimental synthesis of many chemotherapeutic agents is based on the concept of enzymatic inhibition. It is also possible to produce minor alterations in substrates so that the organism generates from it an inhibitor of the metabolic system, a phenomenon called "lethal synthesis."

Competitive Inhibition

From both a theoretical and practical viewpoint the so-called competitive inhibitors are of utmost importance. Competitive inhibitors are structurally similar to the normal substrate. For example, an enzyme called succinate dehydrogenase which catalyzes the conversion of succinate to fumarate is competitively inhibited by malonate, oxaloacetate, and acetoacetate, which are structurally related to succinate.

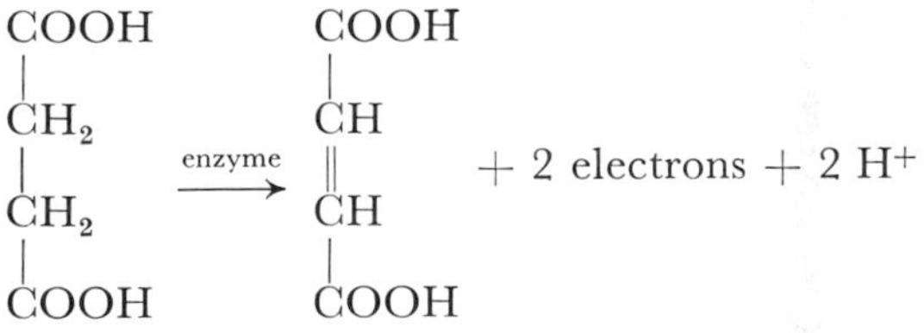

$$
\begin{array}{ccc}
\text{COOH} & & \text{COOH} \\
| & & | \\
\text{CH}_2 & \xrightarrow{\text{enzyme}} & \text{CH} \\
| & & \| \\
\text{CH}_2 & & \text{CH} \\
| & & | \\
\text{COOH} & & \text{COOH}
\end{array}
\quad + 2 \text{ electrons} + 2\text{ H}^+
$$

Succinate   Fumarate

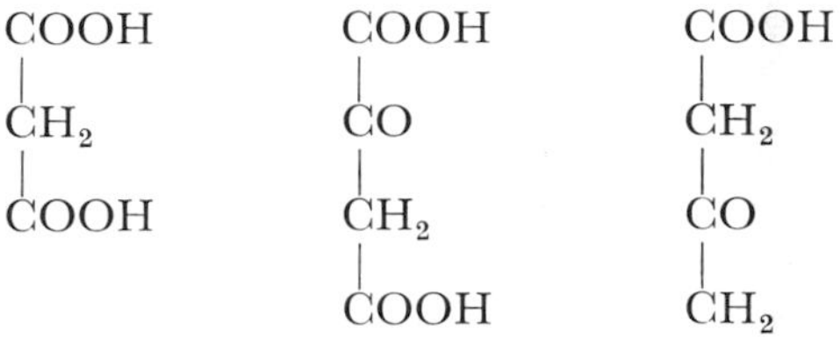

$$
\begin{array}{ccc}
\text{COOH} & \text{COOH} & \text{COOH} \\
| & | & | \\
\text{CH}_2 & \text{CO} & \text{CH}_2 \\
| & | & | \\
\text{COOH} & \text{CH}_2 & \text{CO} \\
 & | & | \\
 & \text{COOH} & \text{CH}_2
\end{array}
$$

Malonate   Oxaloacetate   Acetoacetate

Muscle tissue contains an important contractile protein called myosin which also functions as an enzyme catalyzing the hydrolysis of ATP to ADP.

$$
\text{A—P—P—P} \xrightarrow[\text{Ca}^{++}]{\text{myosin}} \text{A—P—P} + \text{P}_i
$$

Adenosine triphosphate ———→ Adenosine   Inorganic
                                        diphosphate + phosphate

In this case A—P—P and P—P (pyrophosphate) both act as competitive inhibitors of this myosin ATPase.

The basis of competitive inhibition is most probably the fact that many enzymes are not absolutely specific for a substrate and thus a minor alteration of the structure of the substrate may "fool" the enzyme.

In the presence of a competitive inhibitor an enzyme-inhibitor complex similar to the enzyme-substrate complex is formed but this complex is not metabolized further by the enzyme at an appreciable rate; i.e., the formation of the enzyme-inhibitor complex is a "dead end." For this reason such inhibitors are often called dead-end inhibitors. The binding of a competitive inhibitor to the enzyme is reversible and the addition of increased amounts of substrate will either completely or at least to a great extent suppress enzyme-inhibitor complex formation. In competitive inhibition there is an inverse proportionality between the velocity of the reaction and the inhibitor concentration. In the presence of a fixed amount of inhibitor the reaction velocity is proportional to the substrate concentration. In the case of a particular compound it may be important to decide whether or not it acts as a competitive inhibitor. The experimental procedure which is suitable to achieve this goal is based on the following simplified theoretical considerations.

If besides the substrate and enzyme an inhibitor is also present, the following reactions may proceed simultaneously.

Equation 17a:
$$C_E + C_S \underset{k_2}{\overset{k_1}{\rightleftharpoons}} C_{ES} \xrightarrow{k_3} C_E + C_P$$

Equation 17b:
$$C_E + C_I \underset{k_5}{\overset{k_4}{\rightleftharpoons}} C_{EI}$$

C_I = total inhibitor concentration
C_{EI} = total enzyme-inhibitor complex concentration

The Michaelis constants for equations 17a and 17b are the following:

Equation 18a:
$$K_m = \frac{[C_E - (C_{ES} + C_{EI})]C_S}{C_{ES}}$$ for the enzyme-substrate reaction

Equation 18b:
$$K_I = \frac{[C_E - (C_{ES} + C_{EI})]C_I}{C_{EI}}$$ for the enzyme-inhibitor reaction

K_I = dissociation constant of the enzyme-inhibitor complex

Solving equation 18b for C_{EI}:

Equation 19:
$$C_{EI} = \frac{(C_E - C_{ES}) C_I}{K_I + C_I}$$

By combining equations 18a and 19 we obtain:

Equation 20:
$$K_m C_{ES} = C_S \left(C_E - C_{ES} - \frac{C_E C_I - C_{ES} C_I}{K_I + C_I} \right)$$

After rearrangement we obtain:

Equation 21:
$$C_{ES} = \frac{C_E C_S K_I}{K_m K_I + K_m C_I + K_I C_S}$$

It was shown before that:

$$V = k_3 C_{ES} \qquad \text{and} \qquad C_{ES} = \frac{V}{k_3}$$

Equation 22:
$$V = \frac{k_3 C_E C_S K_I}{K_m K_I + K_m C_I + K_I C_I}$$

If the enzyme is saturated with substrate:

$$V_{\max} = k_3 C_{ES} = k_3 C_E$$

Therefore equation 22 may be written as follows:

Equation 23:
$$V = \frac{V_{\max} C_S K_I}{K_m K_I + K_m C_I + K_I C_S}$$

Taking the reciprocals:

Equation 24:
$$\frac{1}{V} = \frac{K_m K_I + K_m C_I + K_I C_S}{V_{\max} C_S K_I}$$

Rearrangement yields:

Equation 25:
$$\frac{1}{V} = \frac{K_m}{V_{\max} C_S} + \frac{K_m C_I}{V_{\max S} C K_I} + \frac{1}{V_{\max}}$$

Equation 26:
$$\frac{1}{V} = \frac{K_m}{V_{\max}} \left(1 + \frac{C_I}{K_I} \right) \frac{1}{C_S} + \frac{1}{V_{\max}}$$

Using equation 26, a plot of $\frac{1}{V}$ against $\frac{1}{C_S}$ in the absence of the inhibitor and in the presence of various inhibitor concentrations will provide valuable information about the inhibitor. A competitive inhibitor does not change the $V_{\max}$ but it does change the K_m of the reaction (Fig. 6-9, curves c and d). The true dissociation constant of the enzyme inhibitor, K_I, may be calculated from the slope of the reciprocal plot in the presence of the inhibitor. The K_m has to be determined separately in the absence of the inhibitor (Fig. 6-9, curve a).

A more potent inhibitor (inhibitor I_2) at equal inhibitor concentrations will show a steeper slope (Fig. 6-9, curve c). This is the consequence of a smaller K_I value. If K_I decreases, $\frac{1}{K_I}$ increases, and $\frac{1}{K_I}$ is the measure of the affinity of the inhibitor to the enzyme.

Noncompetitive Inhibition

In case of "noncompetitive" inhibitor there need not be any structural similarity between the substrate and the inhibitor. The inhibitor need not be bound to the active center and therefore the K_m of the enzyme-substrate complex may remain unchanged. On the other hand, the conversion of the enzyme-substrate complex is suppressed by the inhibitor bound to the enzyme. At a given inhibitor concentration the depressed velocity of the reaction is independent of the substrate concentration.

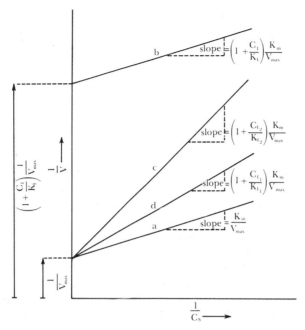

Figure 6-9. Lineweaver-Burk plot of a hypothetical enzyme in the presence and absence of inhibitors. v = Velocity. C_s = Substrate concentration. C_I = Inhibitor concentration. K_m = Michaelis constant. K_I = Dissociation constant of the enzyme-inhibitor complex. V_{max} = Maximal velocity. *Curve a.* No inhibitor was added to the enzyme. *Curve b.* Noncompetitive inhibitor was added to the enzyme. The K_m of the reaction remained unaltered but the maximal velocity is strongly decreased. *Curves c and d.* Competitive inhibitors were added to the enzyme. The K_m of the enzymatic reaction changed in the presence of the inhibitors since the slopes of curves c and d are different from that of curve a. On the other hand, the intercept of these three curves on the y-axis remained the same, denoting that the maximal velocity remained the same in the presence of competitive inhibitors. If C_{I_2} = C_{I_1}, the larger slope of curve c compared to that of curve d indicated that I_2 was a stronger inhibitor than I_1.

This may be explained by assuming that enzyme-inhibitor complex formation is not influenced by the substrate concentration simply because they do not compete for the same protein groups. Virtually every compound capable of forming a stable bond with any functional protein group may become a noncompetitive inhibitor.

Kinetic experiments similar to those discussed in connection with competitive inhibition may reveal noncompetitive inhibition. In the case of noncompetitive inhibition the following reactions proceed simultaneously.

$$C_E + C_S \underset{k_2}{\overset{k_1}{\rightleftharpoons}} C_{ES} \overset{k_3}{\longrightarrow} C_E + C_P$$

$$C_E + C_I \underset{k_5}{\overset{k_4}{\rightleftharpoons}} C_{EI}$$

$$C_{ES} + C_I \underset{k_7}{\overset{k_6}{\rightleftharpoons}} C_{ESI}$$

C_{ESI} = concentration of enzyme-substrate-inhibitor complex

It must be pointed out that the enzyme-inhibitor and the enzyme-substrate-inhibitor complexes are enzymatically inactive. If one plots $\frac{1}{V}$ against $\frac{1}{S}$, a straight line may be obtained, and its intercept with the y-axis is different from that in the absence of the inhibitor.

In this case both slope and intercept of the uninhibited reaction are changed by a factor $\left(1 + \frac{C_I}{K_I}\right)$. It may be recalled that in the case of competitive inhibition only the slope is changed; the intercept remains constant. This offers a rather simple graphical distinction between competitive and noncompetitive inhibitors (Fig. 6-9, curve b).

It should be mentioned, however, that inhibition may behave in a partially competitive and partially noncompetitive manner. More involved kinetic analysis may disclose and define such "mixed-type" inhibitions.

Enzymes are proteins. According to our definitions, any substance or condition which diminishes the velocity of the enzymatic reaction may be considered an inhibitor; therefore if the enzyme protein is denatured by excessive pH, temperature, or other chemical agents, the enzymatic function is usually destroyed. Such denaturation experiments may also provide useful information regarding the correlation of enzymatic function and protein structure in selected cases. The evaluation of such results should always be based upon the nature and extent of the denaturation itself. Occasionally some noncompetitive inhibition (heavy metals, for example) also produces denaturation. Before the results obtained with such "noncompetitive inhibitors" are interpreted, great effort must be made to disclose possible protein structural alterations caused by the inhibition.

Allosteric Inhibition

One interesting and biologically important type of enzyme inhibition was discovered recently. In this case inhibitor and substrate are not bound to the same site on the enzyme, yet the kinetics of this process appears to indicate a competitive type of inhibition. The basic assumption is that an inhibitor bound to groups outside the active site changes the configuration of the latter and depresses the formation of enzyme-substrate complex. Similarly, increased substrate concentration may depress the formation of an enzyme-inhibitor complex at the inhibitor site. The inhibitor site is often called the allosteric site and the process is known as allosteric inhibition (Fig. 6-10).*

The precise kinetic definition of the various types of allosteric inhibitors is beyond the scope of this review. The interested reader should consult a reference book for a technical description.

---

* Allosteric activation of enzyme action is also known. In this case the compound bound to the allosteric site activates rather than inhibits the process at the active site.

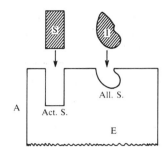

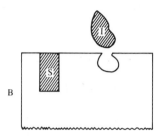

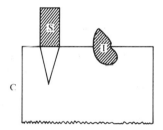

Figure 6-10. Simplified mechanism of action for an allosteric inhibitor. S = Substrate. I = Inhibitor. Act.S. = Active site. All.S. = Allosteric site. *A*, Conformation of the active site and allosteric site in the absence of substrate and inhibitor. *B*, Formation of enzyme-substrate complex altered the conformation of the allosteric site and disproved the possibility for enzyme-inhibitor complex formation. *C*, The enzyme-inhibitor complex formed at the allosteric site induced a change on the active site, inhibiting the enzyme-substrate complex formation.

DIAGNOSTIC ENZYMOLOGY

There is an important medical application of enzymology: quantitative determination of certain enzymes in the diagnosis of disease. The medical diagnostic laboratory does not necessarily compete with the laboratory of the biochemist in purifying and characterizing the individual enzymes. In the diagnostic laboratory most of the investigation involves the enzymatic activity of serum, urine, feces, duodenal juice, and cerebrospinal fluid. In some of these materials the active enzyme concentration is less than 1 per cent of the total proteins present.

The main purpose here is to determine the quantity of a given enzyme in a unit volume of blood, urine, or other specimen. Experience has shown that in certain diseases the concentration of some enzymes in the blood increases, and that of others decreases. The precise cause and mechanism of these changes is not always known but if it is useful for medical diagnosis the enzyme assay is a valuable clinical tool. Most of the enzymes measured are normally of intracellular origin, but some are enzymes normally secreted into the gastrointestinal tract. The normal wear and tear of the various cells of the tissues releases a certain amount of enzymes into the circulation, a most important factor in the maintenance of the normal level of enzymes in blood and urine.

In disease, cells of various tissues may be damaged to the point where the permeability of cell membranes is increased sufficiently to release much of a given enzyme into the blood. On the other hand, tissue responsible for the formation of a certain enzyme may be so damaged by a disease that it cannot form the enzyme, and the concentration of that enzyme will decrease in the blood.

Amylase and Lipase

The determination of the activity of certain digestive enzymes in the serum may provide valuable information about disease processes involving the pancreas or some other digestive gland. If the pancreatic tissue is involved in an acute inflammatory process, the concentration of serum amylase and lipase may be considerably elevated.

Phosphatases

Blood serum is known to possess various phosphomonoesterases. Glucose phosphatase and 5′-nucleotidase are substrate specific but two other enzymes hydrolyze various phosphomonoesters. The latter two, the so-called phosphatases, are more important in the diagnosis of certain diseases. They were discovered because of the differences in their pH optima. The pH optimum of the "acid phosphatase" is around pH 5.0; that of the "alkaline phosphatase" is between pH 8.0 and 10.0, depending upon the substrate used.

It has been established that a major source for acid phosphatase is the prostate. This gland produces roughly 1000 times more acid phosphatase than any other known source. An increased concentration of this enzyme in serum is usually the consequence of bone metastases of prostatic carcinoma. The serum concentration of this enzyme remains high if the therapy is unsuccessful but returns to normal if the tumorous growth has been arrested. Red blood cells also may contain some acid phosphatase, which can easily be differentiated from the prostatic enzyme by using the proper inhibitors (Table 6-1).

The alkaline phosphatase activity in the serum is connected with the osteoblastic activity in the bones. Osteoblasts are involved in bone formation. Infants have more vigorous bone formation than children, and children have a higher rate than adults. Accordingly, the alkaline phosphate activity in blood serum is highest in those up to the age of 2 years (five to ten times the adult value); it decreases somewhat after the second year of life but

Table 6-1. *Differentiation of Erythrocytic Acid Phosphatase from Prostatic Acid Phosphatase*

	0.5% FORMALDEHYDE	0.01 M TARTRATE
Prostatic acid phosphatase	No effect	Inhibition
Erythrocytic acid phosphatase	Inhibition	No effect

during childhood remains about three times as high as the normal adult value. Diseases with increased osteoblastic activity (e.g., Paget's disease and hyperparathyroidism), regardless of their etiology, produce greatly increased alkaline phosphatase activity in the blood. The physiopathological basis for the increased serum alkaline phosphatase activity in connection with obstructive liver disease is less well understood. The best explanation is that this enzyme is excreted by the liver and if there is any mechanical disturbance in the bile flow the alkaline phosphatase can not easily be excreted from the body and thus accumulates in serum. In combination with other tests, serum alkaline phosphatase activity is useful to establish the diagnosis of biliary obstruction.

Transaminases

Plasma transaminase activity has an interesting and important application. Transaminations are important in cellular metabolism and transaminases occur in relatively high concentration in many tissues. For example, the reported distribution of glutamic-oxaloacetic transaminase (GOT) and glutamic-pyruvic transaminase (GPT) is shown in Table 6-2.

It is obvious that for GOT the heart and liver and also the skeletal muscle because of its total size are important sources. As far as GPT is concerned, the liver is the only significant source. Diseases which are connected with cell injury in the liver or cardiac or skeletal muscle may produce high serum levels of GOT, while the simultaneous increase of GPT is almost pathognomonic of liver cell injury.

Enzymatic determinations, like other laboratory tests, must be properly integrated into the general clinical picture, but when used in this way they

Table 6-2. *Distribution of Glutamic-Oxaloacetic Transaminase and Glutamic-Pyruvic Transaminase*

	GOT UNITS/WET TISSUE	GPT UNITS/WET TISSUE
Heart	156.000	7.000
Liver	142.000	44.000
Skeletal muscle	99.000	4.800
Kidney	91.000	19.000
Pancreas	28.000	2.000
Serum	20.000	16.000

provide valuable information. Experience has shown that they provide insight regarding the extent of the cellular damage suffered by the diseased organ and also a guide to the recovery process.

Lactic Dehydrogenase and Isoenzymes

Lactic dehydrogenase (LD) is another widely distributed enzyme. The proportion of LD concentration in the liver, heart muscle, skeletal muscle,

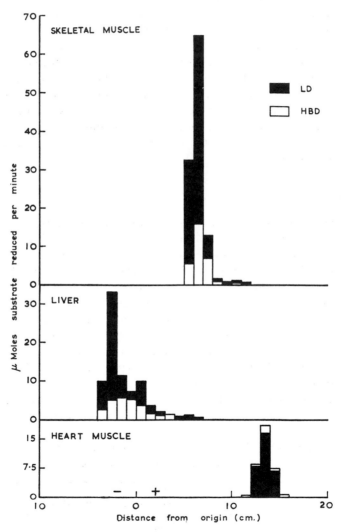

Figure 6-11. Electrophoretic separation of the lactic dehydrogenase isoenzymes. ■ Lactic dehydrogenase activity. □ Ketobutyric dehydrogenase activity. It may be seen that the lactic dehydrogenase enzymes obtained from various tissues have distinctively different electrophoretic mobilities. The ketobutyric dehydrogenase activity of the heart enzyme is higher than its lactic dehydrogenase activity. In the case of the isoenzymes of the liver and muscle, the opposite is true. (From Wilkinson, J. H.: Introduction to Diagnostic Enzymology. Baltimore, The Williams & Wilkins Co., 1962, p. 163.)

and erythrocytes is approximately 5:3:3:1. Under normal conditions its concentration in the serum is about 240 times less than that in the erythrocyte. Since its concentration in several different tissues is high and in the plasma negligible, LD appears to be a suitable enzyme to measure in diagnosis of diseases that damage LD-rich tissues.

With starch gel electrophoresis, serum LD activity was found to migrate in at least three different bands; one band corresponds to the region where γ globulins migrate while the second and third enzymatically active proteins move with the α and β globulins. Further experiments established that the electrophoretically distinct LD enzymes originate from different tissues.

A particular type of enzyme which is synthesized in several different tissues may possess essentially similar catalytic activity regardless of its source. However, the physical and immunological properties of these enzymes may be different and characteristic of the tissue from which the enzyme is derived. Such enzymes are often called isoenzymes. Isoenzymes may be very useful in diagnostic enzymology because they disclose the source of a particular enzyme and indicate the nature of diseased tissue. A further useful difference between the LD isoenzymes is their relative ability to reduce α-ketobutyric acid to α-hydroxybutyric acid.

The LD isoenzymes originating from the heart and skeletal muscle have approximately twice as much ketobutyric reductase activity as the liver LD enzymes (Fig. 6-11).

It is easy to see that the combined information derived from electrophoresis and determination of ketobutyric reductase activity of LD is a great help toward achieving the goals of the medical diagnosis. An increasing number of isoenzymes are now used as tools for medical research and diagnosis.

REFERENCES

1. Cleland, W. W.: Enzyme kinetics. Ann. Rev. Biochem. *36*:77–112 (1967).
2. Dixon, M., and Webb, E.: Enzymes. New York, Academic Press, 1964.
3. Gutfreund, H.: An Introduction to the Study of Enzymes. Oxford, Blackwell Scientific Publications, 1965.
4. Koshland, D. E., Jr., and Kent, K. E.: The catalytic and regulatory properties of enzymes. Ann. Rev. Biochem. *37*:359–410 (1968).
5. Wilkinson, J. H.: An Introduction to Diagnostic Enzymology. Baltimore, The Williams & Wilkins Co., 1962.

NUCLEIC ACIDS

BASES AND OLIGONUCLEOTIDES

Some 100 years ago the name nuclein was given to a phosphorus-containing acidic material extracted from the nuclei of white blood cells. In time it was shown that nucleic acids are major constituents of this material and that all living cells contain nucleic acids. The exact chemical nature was gradually elucidated during the first 60 years of this century, and accompanying the development of this chemical understanding was the unraveling of the biological role of these compounds, a development which turned the mainstream of biological research toward a molecular approach.

Nucleic acids are macromolecules. Macromolecular nucleic acids may be classified according to their sugar component. Deoxyribonucleic acids (DNA) contain deoxyribose; ribonucleic acids (RNA) contain ribose. Macromolecular nucleic acids may be chemically or enzymatically degraded to simple monomeric units called nucleotides. The nucleotides consist of a heterocyclic nitrogen-containing base (a purine or pyrimidine), a pentose sugar (a pentose or deoxypentose), and phosphoric acid. The following heterocyclic nitrogen-containing bases are the most important components of nucleic acids.

Purine Bases

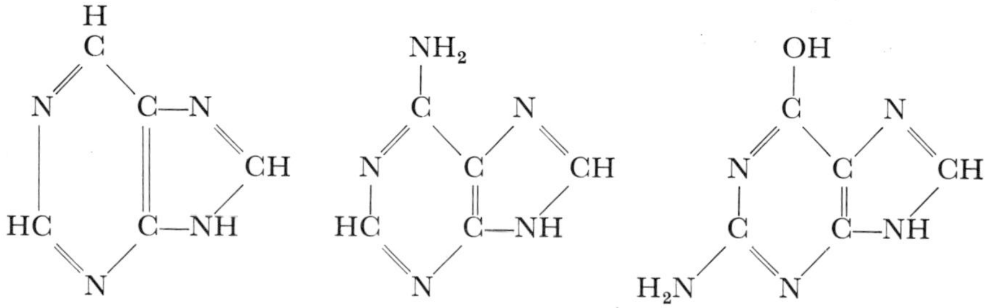

Purine   Adenine   Guanine

Adenine and guanine are the most important purine constituents of nucleic acids. However, other purine derivatives have been isolated from specific nucleic acids such as certain transfer ribonucleic acids (t-rRNA) or viral nucleic acids. The following are examples of some of these.

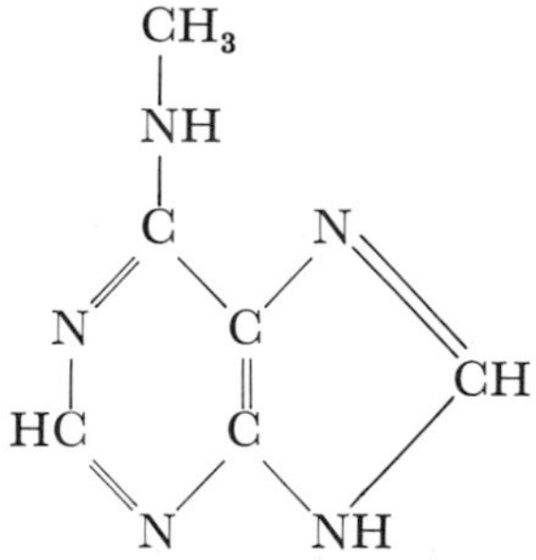

6-Methylaminopurine

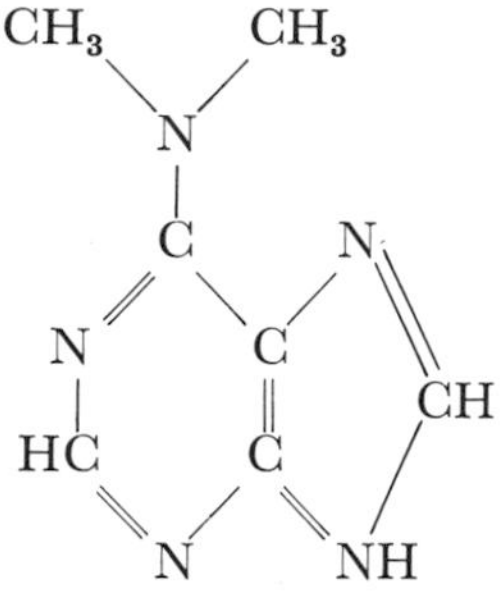

6-Dimethylaminopurine

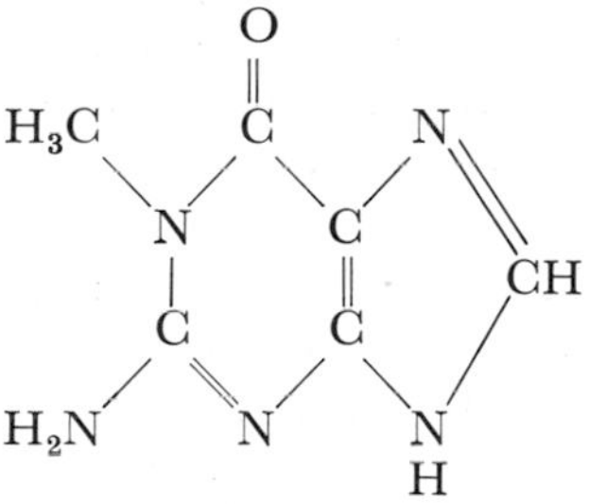

1-Methylguanine

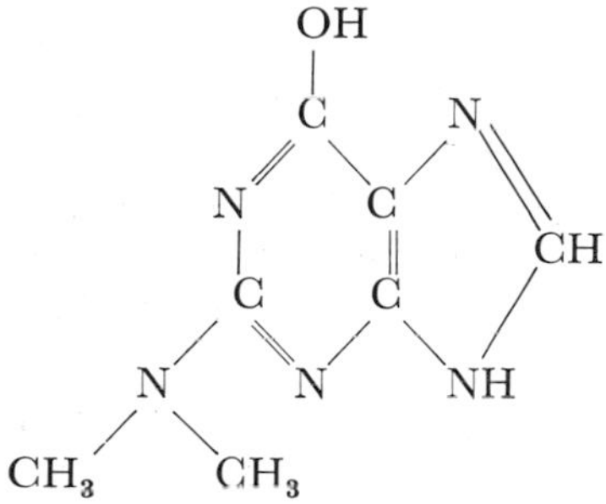

2-Dimethylamino-6-hydroxypurine

Pyrimidine Bases

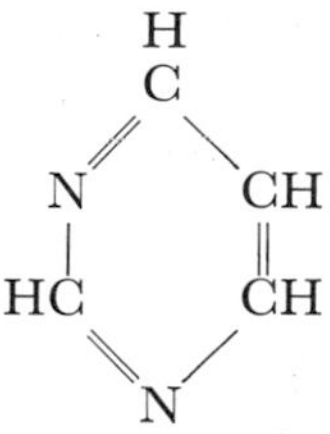

Pyrimidine

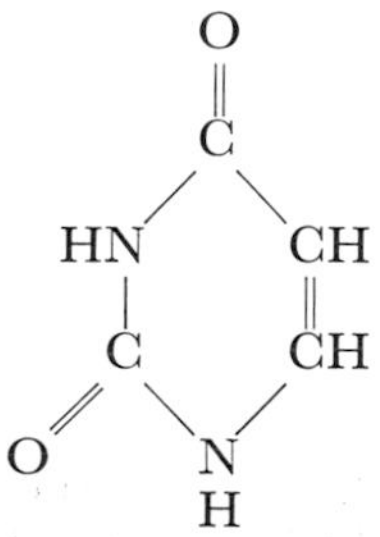

Uracil

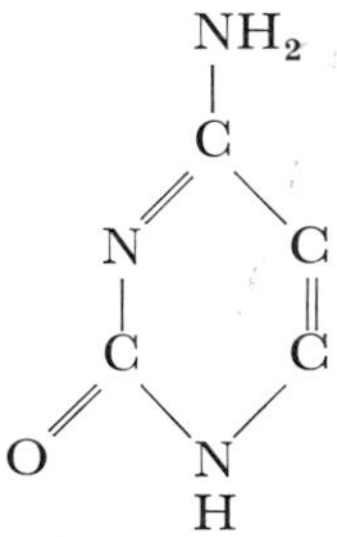

Cytosine

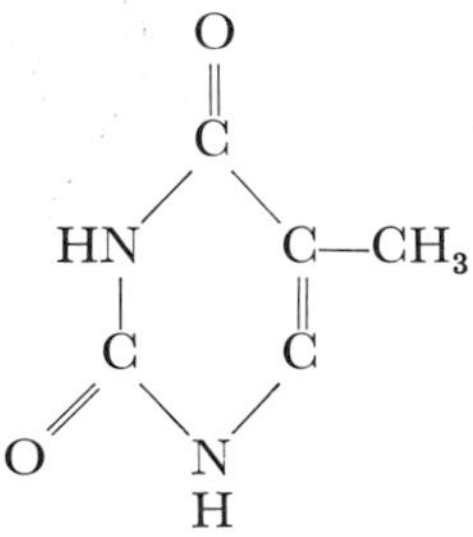

Thymine

Uracil is present only in RNA, thymine is a constituent only of DNA, but cytosine is a constituent of both. In addition to these three major pyrimidines there are other pyrimidines that have been isolated from transfer RNA and viral nucleic acids, e.g., 5-methylcytosine and 5-hydroxymethylcytosine.

5-Methylcytosine   5-Hydroxymethylcytosine

Both purine and pyrimidine bases absorb ultraviolet light, maximum absorption occurring around 260 mμ. The molar extinction coefficients of the various bases differ and the pH of the solution also affects these coefficients. Ultraviolet absorption at 260 mμ is frequently used for detection and quantitation of nucleic acids (Fig. 7-1).

The sugar component of nucleotides is linked to the purine or pyrimidine base by an N-glycoside bond, the combination being called a nucleoside. The pentoses D-ribose and D-deoxyribose are the sugars found in nucleic acids. Deoxyribonucleic acid contains only deoxyribose, whereas D-ribose is the only sugar component of the ribonucleic acids. Figure 7-2 depicts these sugars.

The colorimetric assays of nucleic acids are often based on the estimation of the sugar moiety of the nucleic acids. The orcinol reaction is used to quantitate RNA on the basis of its D-ribose content and the diphenylamine reaction (2-deoxy-D-ribose) serves similarly for DNA assay.

The purine and pyrimidine bases are bonded to the C_1 atom of the

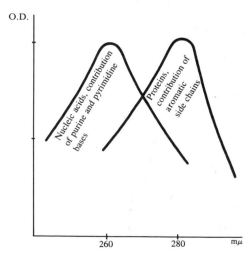

Figure 7-1. Diagrammatic representation of ultraviolet absorption of proteins and nucleic acids at alkaline pH.

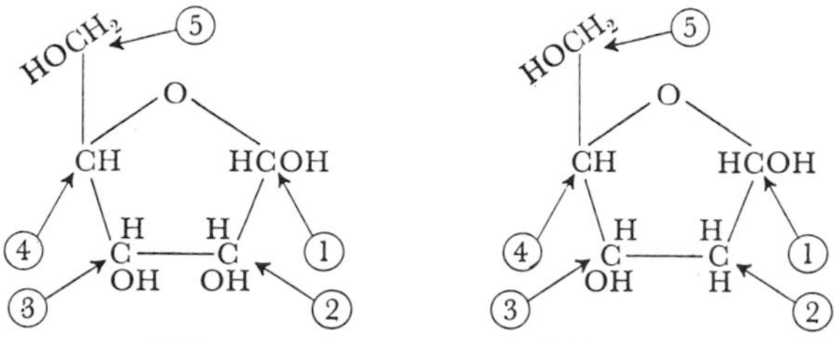

D-Ribose 2-Deoxy-D-ribose

Figure 7-2. The position of carbon atoms in the ribose molecule. Arrows indicate the number of carbon atoms in the ribose molecule. Subsequently we will refer to the ribose carbons in positions 1 to 5 as C_1 to C_5.

pentose in both the DNA and RNA molecules as follows:

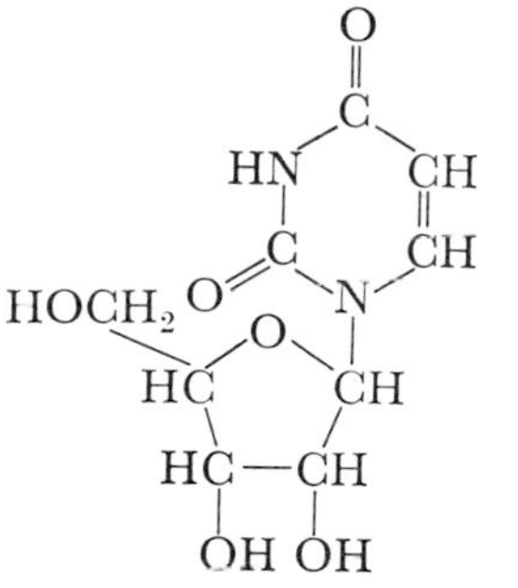

Uridine

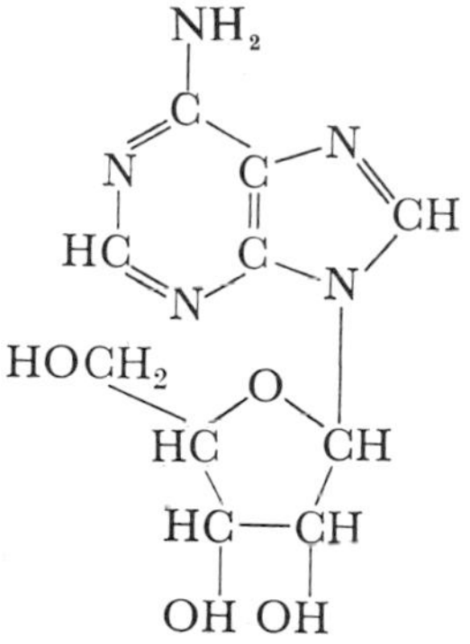

Adenosine

Free nucleotides contain one, two, or three phosphate groups in addition to their nucleoside moiety. The position by which the phosphate ester or phosphate esters are linked to the pentose varies and is indicated by a prefix with a prime sign which refers to the carbon number of the pentose molecule.

Table 7-1. *The Most Important Nucleosides and Corresponding Mononucleotides*

Uracil — D-ribose = uridine
Cytosine — D-ribose = cytidine
Cytosine — 2-deoxy-D-ribose = deoxycytidine
Thymine — 2-deoxy-D-ribose = thymidine
Adenine — D-ribose = adenosine
Adenine — 2-deoxy-D-ribose = deoxyadenosine
Guanine — D-ribose = guanosine
Guanine — 2-deoxy-D-ribose = deoxyguanosine
Uridine — P_i* = uridine monophosphate
Cytidine — P_i = cytidine monophosphate
Deoxycytidine — P_i = deoxycytidine monophosphate
Thymidine — P_i = thymidine monophosphate
Adenosine — P_i = adenosine monophosphate
Guanosine — P_i = guanosine monophosphate
Deoxyguanosine — P_i = deoxyguanosine monophosphate

* P_i = inorganic phosphate.

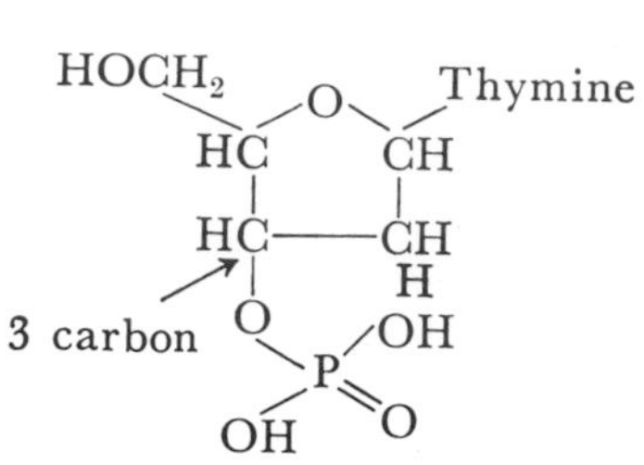

3'-Thymidine monophosphate 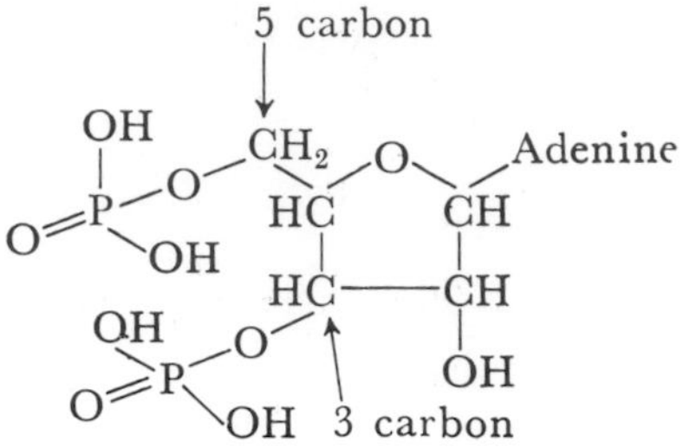   3',5'-Adenosine diphosphate

For example, in the case of 3'-thymidine monophosphate the 3' indicates that C_3 of the deoxy-D-ribose moiety is linked to the phosphate. Another example is 3',5'-adenosine diphosphate, a component of pyridine nucleotide isoenzymes and coenzyme A; in this case the two phosphates are attached to the D-ribose moiety of the adenosine at C_3 and C_5. Adenosine-3',5'-cyclic phosphate (cyclic AMP) is an important mediator of hormone action and has the following structure:

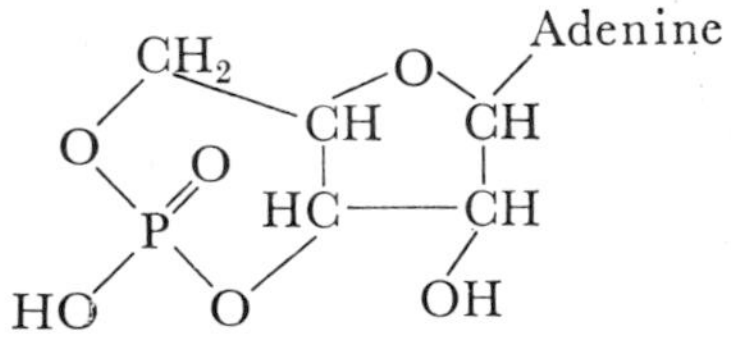

Adenosine-3',5'-cyclic phosphate

Other important adenine nucleotides are adenosine diphosphate (ADP) and adenosine triphosphate (ATP), which have the following structure (note the pyrophosphate linkage):

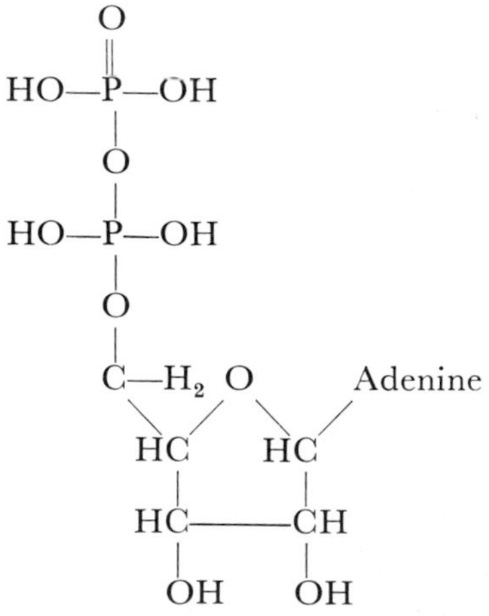

5'-Adenosine diphosphate 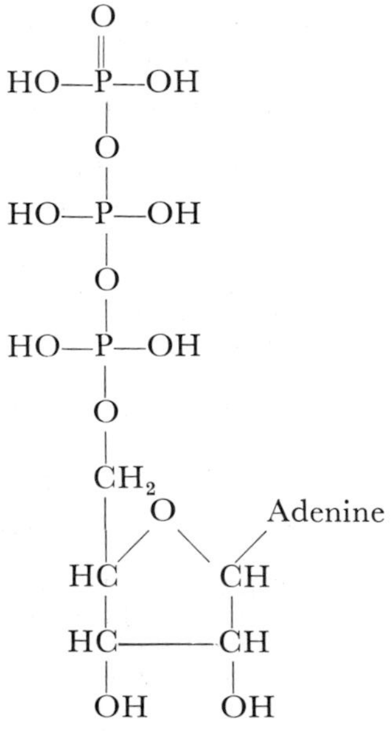   5'-Adenosine triphosphate

THE ·DNA MOLECULE

Biochemical Degradation of the DNA Molecule

Extensive studies on DNA from various sources revealed that the ratio base:D-ribose:phosphate is 1:1:1, indicating that nucleotide monophosphate may be the smallest repeating unit of this macromolecule. A large number of enzymes have been isolated from mammalian, bacterial, and plant sources which promote the hydrolysis of DNA. As a result of such enzymatic action, the viscosity of DNA solutions decreases, the ultraviolet absorption of the solution increases, and small-molecular-weight products appear. All these results stem from the fact that the DNA molecule has been degraded. The enzymes that catalyze the degradation of DNA sequentially by attacking an intranucleotide bond of a terminal nucleotide of the DNA chain are called exonucleases, whereas those enzymes that act on internucleotide links of non-terminal nucleotides within a DNA chain are called endonucleases.

The analysis of the products of enzymatic hydrolysis provides information on the nature of the bonds that hold the backbone of DNA molecule together. It is now established that nucleases catalyze the hydrolytic cleavage of P—O bonds. The simultaneous or sequential action of a deoxyribonuclease (DNAase I) and snake venom phosphodiesterase converts DNA almost completely to 5'-deoxynucleoside monophosphates. DNAase I is a endonuclease which catalyzes the cleavage of intranucleotide bonds between the third carbon atom of the ribose molecule and the phosphorus atom linked to it. Therefore after the action of this enzyme the phosphoric acid will remain linked to the fifth carbon atom of the adjacent nucleoside. Snake venom phosphodiesterase is an exonuclease which splits off a 5'-deoxynucleotide from the end of a DNA chain or from the end of an oligonucleotide. The products of the DNAase I action, the oligodeoxynucleotides, serve as substrates for the snake venom phosphodiesterase, which completes the hydrolysis of the DNA molecule to yield 5'-deoxynucleoside monophosphates (see Fig. 7-3).

Similarly, the combined hydrolysis of DNA by another DNAase (DNAase II) and a spleen phosphodiesterase yields 3'-deoxynucleoside monophosphates. DNAase II is an endonuclease which cleaves the intranucleotide bonds between the fifth carbon atom of the ribose molecule and the phosphorus atom. After the action of this enzyme the phosphoric acid remains attached to the third carbon atom of the adjacent nucleoside. Spleen phosphodiesterase is an exonuclease which splits off a 3'-deoxynucleotide from the end of a DNA chain or from the end of an oligonucleotide. Like snake venom phosphodiesterase, spleen phosphodiesterase also uses the oligodeoxynucleotides as substrates. This enzyme completes the action of DNAase II on the DNA molecule (see Fig. 7-3).

These results, together with the results of chemical DNA degradation, strongly support the view that in the DNA molecule the nucleoside units are

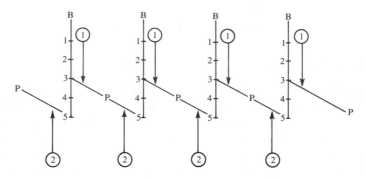

Figure 7-3. Schematic representation of the bonds attacked by phosphodiesterases. If a phosphodiesterase splits all the bonds indicated by arrows 1, 5′-nucleoside monophosphates are liberated, while other phosphodiesterases which attack DNA at the bonds shown by arrows 2 produce 3′-nucleoside monophosphates after complete hydrolysis. An endonuclease splits some of these bonds within the "body" of the DNA molecule and liberates oligodeoxyribonucleotides. An exonuclease splits some of these bonds at the free ends of the molecule and is capable of producing mononucleotides as the end result of a prolonged hydrolysis.

linked together by 3′—5′ phosphodiester linkages (see Fig. 7-3). It was already established that the N-glycosidic linkage between the base and D-ribose involves the C_1 atom on the ribose. C_2 is fully hydrogenated and C_4 in the furanose form does not possess a OH group; therefore only C_3 and C_5 of the D-ribose can participate in the phosphoester formation. The participation of both these groups is now firmly established.

Chargaff Rules

As a result of the investigations of Chargaff and his collaborators certain generalizations about the base composition of DNA can be drawn:

1. $A + G = T + C.$* The sum of purines base is equal to the sum of pyrimidines in the DNA molecule.

2. $A = T$ and $G = C$. The molar quantity of adenine is equal to that of thymine and the molar quantity of guanine is equal to that of cytosine.

3. $A + C = G + T$. The number of 6-amino bases is equal to the number of 6-keto bases.

These rules (Chargaff's rules) are the analytical basis for the Watson-Crick model of DNA molecule.

4. The base composition of DNA may be expressed as the mean guanidine + cytidine content. This is usually defined as $\dfrac{G + C}{G + C + A + T}$ which is similar in the various tissues of an organism. On the other hand, the same expression gives significantly different values in various species (see Table 7-2).

---

* A = adenine; C = cytosine; U = uracil (in RNA); G = guanine; T = thymine.

Table 7-2. *Guanidine + Cytidine Content of Various Species*

SPECIES	$\dfrac{G + C}{G + C + A + T}$
Various bacteria	0.25–0.75
Fungi and algae	0.35–0.66
Invertebrates	0.34–0.44
Vertebrates	0.40–0.44

5. Since $A + G + T + C = 1$, any significant deviation from unity should disclose the presence of some "unusual" bases. Such "unusual" bases have been demonstrated in viruses.

The Watson-Crick Model of the DNA Molecule

X-ray diffraction patterns of DNA show the presence of helical poly-nucleotide chains in which the bases are arranged like a "pile of pennies." The importance of the x-ray diffraction findings is emphasized by the fact that DNA molecules obtained from a wide variety of tissues and species give similar patterns.

If DNA solutions at pH 7.0 are titrated with acid to pH 2.0 and then with alkali to pH 7.0, or alternatively DNA is titrated from pH 7.0 with alkali to pH 12.0 and back to 7.0 with acid, the curves obtained from pH 7.0 to alkaline or acid pH differs markedly in shape from those obtained when titrating to a neutral pH. This phenomenon is believed to be due to an irreversible change in DNA structure caused by an acid or alkaline environment. A good explanation is that during the acid or alkali titration hydrogen bonds are broken, which alters the helical structure, and that these bonds are unable to re-form during the back titration and therefore the titration and respective back titration curves do not match.

Since $A = T$ and $G = C$, Watson and Crick proposed that in the DNA molecule hydrogen bond formation between bases is possible only if adenine is paired with thymine and guanine with cytosine (Fig. 7-4).

On the basis of x-ray diffraction data, Chargaff rules, and the concept of hydrogen bonding between the base pairs, Watson and Crick proposed a model for the DNA molecule. This model accounts for the most important physical and chemical characteristics of the DNA molecule and provides a firm basis for the understanding of the biological mechanism of DNA action. According to Watson and Crick, the DNA molecule consists of two matching polynucleotide chains (Figs. 7-5 and 7-6). The two polynucleotides run spirally around a common axis, forming two single helical chains. Both chains are right-handed spirals but they run in opposite directions. (The direction of a chain is determined by the sequence of the components of the phosphodiester linkages—C_3—O—P—O—C_5, the opposite being C_5—O—P—O—C_3 direction.) The backbone of the individual chains is composed of phosphodiester linkages and the two chains are held together by

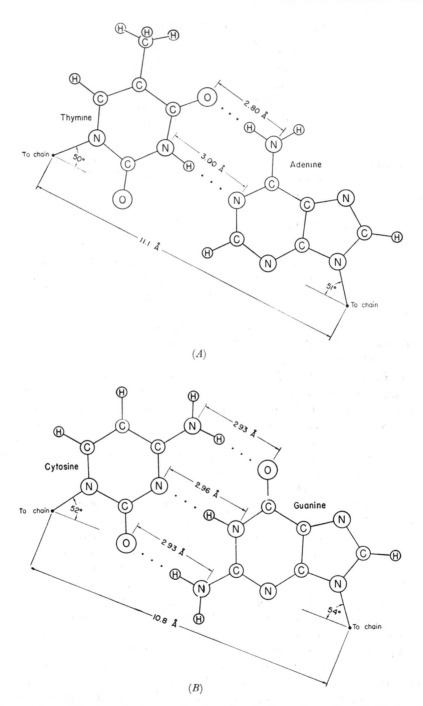

Figure 7-4.   *A*, Hydrogen-bond formation between adenine and thymine. *B*, Hydrogen bond formation between cytosine and guanine. (From Pauling, L.: The Nature of the Chemical Bond. Ithaca, Cornell University Press, 1960, p. 502.)

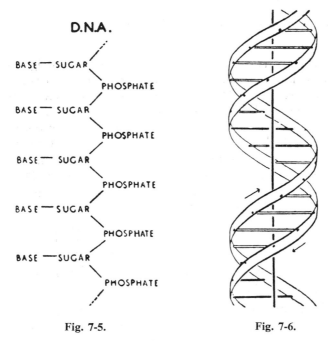

Fig. 7-5.   **Fig. 7-6.**

Figure 7-5. Diagrammatic representation of the DNA molecule. Chemical formula of a single chain of deoxyribonucleic acid. (From Watson, J. D., and Crick, F. H. C.: Nature *171*:964, 1953.)

Figure 7-6. Diagrammatic representation of the DNA molecule. This figure is purely diagrammatic. The two ribbons symbolize the two phosphate-sugar chains, and the horizontal rods the pairs of bases holding the chains together. The vertical line marks the fiber axis. (From Watson, J. D., and Crick, F. H. C.: Nature *171*:964, 1953.)

hydrogen bonds between the matching bases of the two chains. The bases are inside the helix in a position perpendicular to the longitudinal axis of DNA molecule. X-ray diffraction data show the distance between two base pairs to be about 3.4 Å. There are ten bases per turn of the double helix.

Hydrodynamic and optical methods indicate that the molecular weight of the DNA molecules ranges around 5 to 12 $\times$ 10^6 and that the shape of the molecule is that of a stiff coil. The average length of a DNA molecule may be approximately calculated by the following equation:

$$\text{Distance between nucleotide pairs} \times \text{Number of nucleotide pairs} = \text{Length of DNA molecule}$$

$$\text{Number of nucleotide pairs} = \frac{\text{M.W. of DNA}}{\text{M.W. of nucleotide pairs}}$$

Therefore:

$$\text{Distance between nucleotide pairs} \times \frac{\text{M.W. of DNA}}{\text{M.W. of nucleotide pairs}} = \text{Length of DNA molecule}$$

Numerically:

$$3.4 \text{ Å} \times \frac{6 \times 10^6 \text{ gm.}}{2 \times 330 \text{ gm.}} = 30,000 \text{ Å} = 3\mu$$

Some Physicochemical Properties of the DNA Molecule

There is some uncertainty regarding the molecular weight and length of the DNA molecules because of their extreme sensitivity to shearing forces. Unfortunately exposure of the molecule to shearing forces is unavoidable during extraction and purification. DNA extracted from various sources always consists of a heterogeneous population of molecules but it is difficult to know if the DNA of a cell is of heterogeneous structure or whether such heterogeneity was induced during extraction and purification. Both DNA and RNA absorb ultraviolet light in wavelength range of 260 mμ. Both native DNA and native RNA, however, absorb less at 260 mμ than would be predicted on basis of their nucleotide content.

$$\text{O.D. of } \begin{pmatrix} [A]_{n1} \\ [G]_{n2} \\ [T]_{n3} \\ [C]_{n4} \end{pmatrix} \text{ DNA molecule} \quad < \quad \text{O.D. of } \begin{pmatrix} [A]_{n1} \\ [G]_{n2} \\ [T]_{n3} \\ [C]_{n4} \end{pmatrix} \text{ Free bases}$$

O.D. = optical density

In the case of DNA the O.D. at 260 mμ is about 40 per cent less than the sum of the O.D.'s of its constituent bases. This marked hypochromicity is probably a consequence of dipole-dipole interactions between the closely stacked chromophores (bases) within the double helical structure of the native DNA molecule. The extent of hypochromicity appears to depend on the A + T concentration of this molecule. It is generally higher with DNA molecules than with RNA molecules. Double-stranded DNA and RNA have higher hypochromicity than the respective single-stranded forms.

For these reasons hypochromicity has become an important tool for the study of the secondary structure* of nucleic acids. Unfolding, denaturation, or any decreased helical structure of nucleic acids involves an increase in their O.D. at 260 mμ. Conversely, renaturation, or formation of an increased order in the nucleic acid structure, involves a decreased O.D. at 260 mμ.

Extensive studies aimed at elucidating the relative importance of various chemical bonds in maintaining the secondary structure of the DNA molecule have been conducted. Using various solvents it has been established that the stability of the DNA molecule is dependent upon electrostatic, van der Waals, and hydrophobic forces and hydrogen bonds. The phosphate residues in the backbone of the polynucleotide chains carry strong negative charges which repel each other. The addition of cations reduces these repelling electrostatic forces to a considerable extent. Some bases are poorly soluble in water and therefore are rather hydrophobic compounds and are highly packed in the interior of the double helix; it is likely that hydrophobic bond formation occurs between such adjacent bases. These are the hydrophobic bonds which are believed to be important in the stabilizing of the structure of

---

* The secondary structure in the case of the nucleic acid carries the same general meaning as in the case of the proteins.

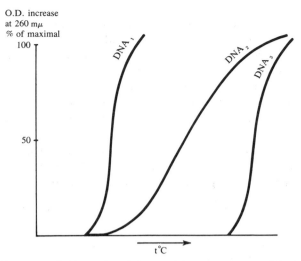

Figure 7-7. Diagrammatic illustration of the transition temperatures of the three hypothetical DNA solutions. DNA_1 has a sharp temperature transition at a relatively low temperature. It has low C + G content and appears to be homogeneous as judged by the criterion of temperature transition. DNA_3 also has a sharp temperature transition at high temperature. This DNA contains C + G in large concentration and appears to be homogeneous. DNA_2 is a mixture of DNA_1 and DNA_3. It has a wide temperature transition range, indicating molecular heterogeneity.

DNA. DNA molecules are denatured by organic solvents, urea, highly acidic or alkaline solutions, low-ionic-strength solutions, ultraviolet irradiation, and heating. In regard to heating it is established that the so-called transition temperature of a DNA (transition from ordered to disordered state) relates to the C + G content of the molecule. Figure 7-7 shows three hypothetical DNA-containing solutions (DNA_1, DNA_2, and DNA_3) in which denaturation is assessed by the disappearance of hypochromicity at 260 mμ. DNA_1 and DNA_3 exhibit a narrow and DNA_2 a wide transition temperature range.

Such an increase in the range of transition temperature as seen in the case of DNA_2 may be due to the heterogeneity of the DNA solution investigated. Those DNA molecules that have a lower C + G content are denatured at a lower temperature while others having higher C + G content need a higher temperature to lose their hypochromicity. Both DNA_1 and DNA_3 show a narrow transition temperature range but at a different temperature. Since C and G are linked by three hydrogen bonds and T and A by only two such linkages, the higher transition temperature needed for the C + G-rich DNA molecules may indicate the participation of hydrogen bonds in the structural stability of the DNA molecule.

Effects of Some Physical and Chemical Agents on the DNA Molecule

Since DNA molecules store and transmit the genetic information in cells, a structural change of these molecules may cause genetic alterations in the species. In Table 7-3 is a compilation of the effects of several physical and chemical agents on the DNA molecule. Most agents listed have manifold effects rather than a single effect on DNA molecules and therefore biological changes caused by such agents cannot be attributed unequivocally to a unique action on the DNA molecule. Nevertheless the use of physical and chemical mutagenic agents does yield some information about the relationship between changes in the molecular structure of the DNA and the genetic modifications.

Similarly, chemical modifications of some base rings may alter the base pairing in certain areas of the DNA molecule. Such alterations often lead to alterations in some genetically controlled quality of the species. The

Table 7-3.   *Effects of Various Physical and Chemical Agents on DNA*

	RUPTURE OF INTERNUCLEOTIDE PHOSPHODIESTER LINKAGES	RUPTURE OF COVALENT BONDS WITHIN THE NUCLEOTIDES	RUPTURE OF HYDROGEN BONDS	FORMATION OF NEW CROSS-LINKING COVALENT BONDS
Ionizing radiation	Yes	Only in high doses	Yes	Cross-links different DNA molecules
Sonic and ultra-sonic waves	Yes	—	—	—
Shearing forces	Yes	—	—	—
Specific enzymes	Yes	—	—	—
Heating	Only after a long time and at high temperature	—	—	—
Alkylating agents	Subsequent to the attack on base rings. Yes	Attacks: N_7 of guanine N_1 and N_3 of adenine N_1 of cytosine	Indirectly yes	Bifunctional reagents may cross-link guanines, using the N_7 atoms on the bases
Strong acid or alkali	Only under drastic conditions	Only under drastic conditions	Yes	—
Succinyl peroxide	Only to a small extent	Both purine and pyrimidine rings oxidized	To a small extent	—
Nitrous acid	Only to a small extent	Replaces free NH_2 groups of bases with OH group	Yes	May introduce new cross-links between the two chains of the DNA
Hydroxylamine	—	Interacts with cytosine	Only to a small extent	—
Urea and guanidine	—	—	Yes	—
Ultraviolet irridation	—	Only in high doses	Indirectly yes	At 260 mμ adjacent thymines may be cross-linked

formation of new covalent bonds between the two chains of the DNA molecule or between different DNA molecules may inhibit the separation or proper recombination of the polynucleotide chains of the DNA molecule and therefore replication of such cells may be suppressed. It may be seen from Table 7-3, however, that most cross-linking agents also interact with the bases so that the observed effects many times cannot be unequivocally ascribed to the formation of new cross-links.

Chemical alterations which produce deletion of some bases or fission of some internucleotide linkages may be replaced or repaired improperly by the cell. The comparison of the structures of the original and "wrongly repaired" DNA and the biological mutation connected with this process may yield interesting results which eventually will lead to the narrowing of the gap in our knowledge concerning the correlation between the structural change of DNA and the genetic event.

THE RNA MOLECULES

Classification of RNA Molecules

RNA is found in the cytoplasm of cells and, in much smaller concentration, also in the nucleus, especially in the nucleolus. RNA can be classified into three subfamilies according to the molecular weight and biological half-life: (1) macromolecular RNA (ribosomal, viral RNA; long half-life); (2) messenger RNA (short half-life); and (3) soluble RNA (transfer RNA; lower molecular weight).

Physical and Chemical Properties of Macromolecular RNA

Macromolecular RNA accounts for 50 to 80 per cent of the total cellular RNA. It is present in the cytoplasm as a component of ribonucleoprotein particles called ribosomes. In many cells large aggregates of ribosomes, often called polysomes, are attached to the endoplasmic reticulum membrane system.

The molecular properties of ribosomal RNA are very similar to those of viral RNA. The ribosomal particles, containing RNA and protein, are characterized in terms of an approximate sedimentation coefficient. Using this approach, two classes of ribosomal particles are found, one with 30S and the other with 50S sedimentation coefficient. Under certain proper experimental conditions these fractions may be aggregated and sediment as 70S and 100S particles. With a phenol extraction procedure the RNA can be isolated from these particles. Usually the molecular weight of this RNA ranges from 5 to 10×10^6. Physical-chemical studies using ultracentrifugation and viscosity measurements indicate that RNA obtained from ribosomal or viral sources behave as highly flexible molecules. At higher ionic strength (>0.2 μ) or lower temperatures ($0°$ to $35°C$.) they behave hydrodynamically as tightly coiled and contracted single chains, while at low ionic strength (<0.1 μ) or higher temperature ($>50°C$.) they display the properties of a random coil.

X-ray diffraction measurements disclose the presence of helical structures within the RNA molecule. Titration curves with acid and alkali and subsequent return to neutrality indicate that an irreversible change in the structure, possibly due to disruption of hydrogen bonds, occurs in the case of the RNA molecules but to a smaller extent than with DNA.

Hypochromicity is also observed with RNA but again to a smaller extent than with DNA. The hypochromicity as well as the optical rotatory dispersion measurements indicate that a fairly large proportion of the molecule exists in helical form when in a solution of high ionic strength at low temperatures. Lowering the salt concentration or raising the temperature decreases the helical content of these RNA molecules (Fig. 7-8).

Analysis of the base composition of RNA molecules indicates that Chargaff rules do not apply to these molecules with the exception that the number of 6-amino compounds is approximately the same as the number of 6-keto compounds, $A + C = G + U$.

It is therefore believed that single-stranded RNA molecules possess a high helical content which is stabilized by electrostatic forces (salt effect), hydrophobic interaction between closely stacked bases (hypochromicity), and hydrogen bonds between matching bases (hysteresis of titration curves). Since the base composition is such that in most RNA molecules a certain number of bases are not matched for hydrogen bond formation, their

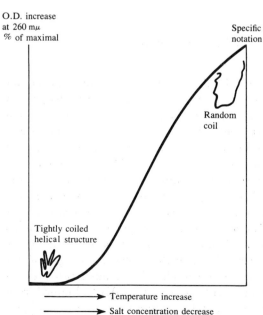

Figure 7-8. Helix-coil transition of a hypothetical RNA molecule as a function of temperature and salt concentration of the mixture. The helix-coil transition was measured by the increase of optical rotation at the sodium D line and by optical density increase at 260 mμ.

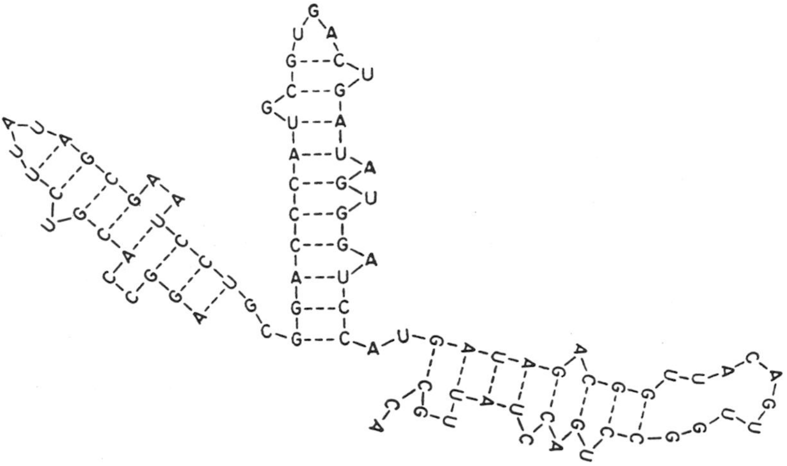

Figure 7-9. Possible base-pairing within singly stranded macromolecular RNA. Helical regions are represented by parallel lines of bases joined by dashed lines, for hydrogen bonds. Bases offset from these lines represent loops outside the helices. (From Fresco, J. R., et al.: Nature *188:*98, 1960.)

participation in the RNA chains is depicted as resulting in loops turning outward from the main helical chain of the molecule (Fig. 7-9). This looping out of a certain number of bases results in a less stable structure than when the bases are closely packed and hydrogen bonded. Although the base sequence analysis of these giant molecules is not completely resolved, nearest-neighbor analysis (see nucleic acid synthesis, Chapter 10) provides a good indication that it is to a great extent ordered. The loci of the hairpin turns may be arranged in such a way that the number of the nucleotides that are looped out is kept at a minimum and the number of possible hydrogen bond formations between matching pairs of bases is kept at a high level. This way the RNA molecule may gain high helical content and high structural stability.

Messenger RNA

Messenger RNA (mRNA) is a biologically important fraction of the RNA because the genetic message is transmitted from the DNA molecules to the mRNA molecule, which in turn carries the message from the nucleus to the site of protein synthesis in the cytoplasm. Messenger RNA accounts for about 4 per cent of the total RNA in cells. Its short half-life and the small quantity are probably responsible for the quite limited amount of information on the molecular characteristics of this RNA fraction. The molecular weight ranges between 1×10^5 and 1×10^6. Sedimentation studies reveal considerable heterogeneity of the mRNA molecules.

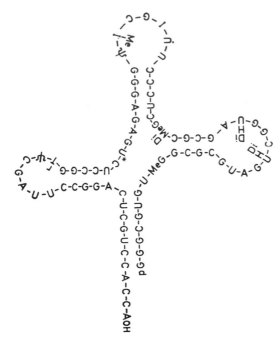

Figure 7-10. The "clover leaf" structure of alanine transfer RNA. The anticodon region is "looped out" from the molecule and is located on the top of this figure. I—G—C (inosine–guanine–cytosine) is the anticodon triplet (for the structural formula of inosine see Chapter 10). With respect to the complementary base pairing inosine seems to behave similarly to guanine. The C—C—AOH sequence on the bottom of the figure is the 3′ terminal region of the molecule which is believed to be site for combination with alanine. (From Holley, R. W. et al.: Science *147*:1462, 1965.)

Transfer RNA (Soluble RNA)

This RNA fraction does not sediment in the ultracentrifuge even after several hours of centrifugation at 100,000 × g. The role of transfer RNA (trRNA) in the protein synthesis is firmly established. It participates in the translation of the genetic message from the base sequence of mRNA to the amino acid sequence of the newly synthesized polypeptide chain. Generally this fraction constitutes about 20 per cent of the total cellular RNA. Several different amino acid trRNA molecules have been isolated. The molecular weight of such molecules ranges from 20,000 to 30,000. Each molecule appears to be homogeneous with a base composition distinctly different from that of ribosomal RNA. One interesting quality of this RNA fraction is the presence of unusual bases (see earlier in this chapter). The complete structure of alanine trRNA has been worked out by Holley and his collaborators. According to these workers, trRNA probably has a configuration similar to a cloverleaf (Fig. 7-10).

The so-called anticodon region (nodoc)* of the trRNA must come close to the mRNA in order to attach its anticodon triplet to the codon triplet of the mRNA. It may be seen from Figure 7-10 that the anticodon triplet is looped out from main chain of the trRNA and the three bases involved in the

---

* Codon triplet is the critical sequence of three bases on the mRNA molecule which carries the genetic message from the nuclear DNA for the protein synthesis. Nodoc is the proper "receptor" triplet on the trRNA for the codon triplet.

codon-anticodon interaction are free for base pairing with the matching bases on the mRNA. The 3' terminal region of the trRNA is the place for the combination with the specific amino acid. Transfer RNA's specific for alanine, serine, tyrosine, valine, leucine, isoleucine, and so forth have been isolated and their base sequence is under intensive investigation.

THE NUCLEOPROTEINS

The DNA of the cell nucleus is tightly associated with protein. The DNA-protein complex appears to be importantly involved in the structural integrity of the nuclear DNA and to be needed for the proper arrangement of the chromosome. Indeed the nature of the DNA-protein complexes appears to have a profound influence on the transference of the genetic information stored in the DNA molecules.

Protamines

Protamines, small (M.W. 4.000 to 10.000) basic proteins, are often the protein components associated with the DNA. About two-thirds of the amino acids of protamines possess positive charges at pH 7.0. Most of these positively charged amino acids are arginine. Deoxyribonucleoprotamines isolated from various sources usually contain 70 per cent DNA and 30 per cent protamine. This ratio if expressed in terms of the negative charges (P_i^-) of DNA and positive charges (arginine$^+$) of protamine results in a neutral molecule. The deoxynucleoproteins are dissociated by 2 M NaCl solutions and it has been shown that in the dissociated form the number of arginine residues is equal to the number of free phosphate residues.

Since the protamine is a much smaller molecule than DNA, many protamine molecules interact with one DNA molecule. Since the phosphate molecules participate in the backbone structure of DNA and the bases are inside the helical structure, the phosphate-arginine electrostatic links are found on the outside of the DNA double helix (Fig. 7-11). Therefore protamine forms the outside coat of the deoxyribonucleoprotein molecule.

Histones

Histones are larger (M.W. 10,000 to 20,000) and somewhat less basic proteins than the protamines. About one-third or less of their total amino acids carry positive charges at neutral pH. "Histones," like protamines, are frequently found in close association with DNA. The deoxynucleohistones isolated from thymus gland have a molecular weight of 1.6 to 20 $\times$ 10^6. DNA, which is present in its double helical form, constitutes about 50 per cent (weight basis) of the nucleohistone. The other 50 per cent is comprised of histone molecules. There is some experimental evidence that DNA molecules

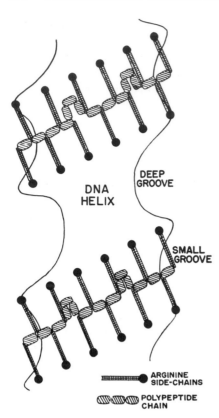

DNA
HELIX

DEEP
GROOVE

SMALL
GROOVE

ARGININE
SIDE-CHAINS

POLYPEPTIDE
CHAIN

Figure 7-11. Diagram showing how prot-amine binds to DNA. The polypeptide chain winds around the small groove on the DNA helix. The phosphate groups are at the black circles and coincide with the basic ends of the arginine side chains. Nonbasic residues are shown in pairs at folds in the polypeptide chain. (From Wilkins, M. F.: Cold Spring Harbor Symp. Quant. Biol. *21*:83, 1956.)

which are fully complexed with histones are inactive as primers for RNA synthesis. It has been proposed that histones regulate gene function.

Viruses

Viruses, which are believed to be the smallest entities exhibiting some lifelike characteristics, consist of DNA or RNA plus protein. It has been shown conclusively that in the case of the RNA-viruses, the RNA molecule and not the protein or any DNA contamination is the material which stores and transmits genetic information. One of the RNA-viruses, the tobacco mosaic virus (TMV) has been investigated in great detail, with respect to the correlation between its chemical structure and biological function. TMV is a ribonucleoprotein (see Fig. 7-12) which shows the arrangement of the RNA in the protein coat. The RNA molecule (M.W. 2.5×10^6), present as a single helical chain, runs in the inside of a protein (M.W. 100.000). About 350 protein molecules (2100 subunits) are needed to surround the RNA molecule completely. When RNA and protein are separated and the infectivity of the two components is investigated, the TMV protein alone is found to have no infectivity at all. On the other hand, the isolated TMV-RNA displays a low infectivity. The infectivity is abolished by RNAase

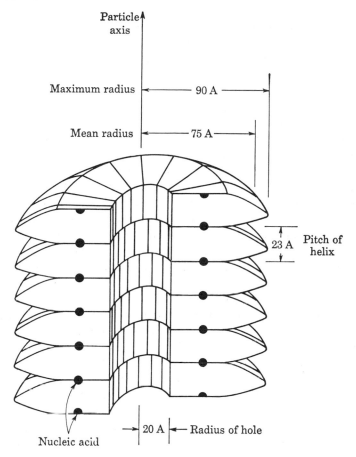

Figure 7-12. Cross-section a short distance along the length of the rod of tobacco mosaic virus. (From Franklin, R. E., Klug, A., and Holmes, K. C.: *In* The Nature of Viruses. Ciba Foundation, London, Churchill Ltd., 1957, p. 39.)

but not by DNAase or proteolytic enzymes. Clearly, therefore, RNA contains the genetic information necessary to replicate the virus and also transmits this information to the host cell. The mechanism for the replication of the single RNA cell has not been elucidated.

PROTEIN SYNTHESIS

The identity and function of a mature cell are determined by its enzyme content and its structural characteristics. The enzymes are exclusively proteins and cellular structure may also depend to a great extent on specific proteins. Since unicellular and multicellular organisms are able to reproduce themselves there must be some material in which the information for the reproduction of the specific proteins is stored within the cells. Observations with light microscopy have shown that during cell division (mitosis) the nucleus of the cell splits first. The nuclear division has already been completed when the division of the protoplasm begins.

The Role of Nuclear DNA

The bulk of the nuclear material within cells gives strong color reaction with some dyes and is called chromosomes. Functionally the chromosome was thought to consist of large numbers of "unit hereditary factors" which contain and store the genetic information to control the structural and functional identity of the cell or organism. Advancement in cellular fractionation made it possible to determine the chemical composition of the cellular nucleus. DNA was shown to be present in large concentrations, together with proteins. Thus the morphological and genetic concept of cellular nucleus and chromosome obtained a firmer chemical basis.

It was established that chromosomes consist of large amounts of DNA, very small amounts of RNA, proteins, and divalent cations. Comparison of the chemical composition of the chromosomes obtained from higher organisms with that of chromosomes isolated from fish made it evident that DNA is the only common chemical component. While great variations were discovered in the base composition of DNA's obtained from different organisms, the DNA composition of the various cells within a given organism was shown to be very similar. Experiments with various physical and chemical "mutaganic" agents showed that alteration in the DNA structure of the germ cells may lead to altered qualities in the progeny.

When some viruses infect cells it is only their DNA (or RNA) which enters the host cell. But this DNA or RNA alone is enough to transmit the genetic information of the virus to the host cell and lead to the production of viral proteins, that is, to the reproduction of the virus itself within the host cell. Evidently every other important component for the viral protein synthesis is available within the host cell. Viral DNA or RNA was necessary, however, and sufficient to "reprogram" the machinery within the host cell.

Sanger's brilliant work on the amino acid sequence of insulin dispelled any doubt that the sequence of amino acid within a given protein is strictly defined. Therefore any theory of protein synthesis must be based on such experimental evidence which accounts for the reproduction of the specific amino acid sequence for any given protein synthesized. For this reason the assumption that templates play an important role in this process became widely accepted.

In mammalian organisms the genetic information, therefore, is stored and transmitted from parents to progeny by the DNA molecules of the nucleus. The base composition and steric configuration of these molecules are mainly responsible for the specificity of the message they store. Understanding of the nature and sequence of events in transmitting the genetic message first became possible when the chemical process of DNA division and

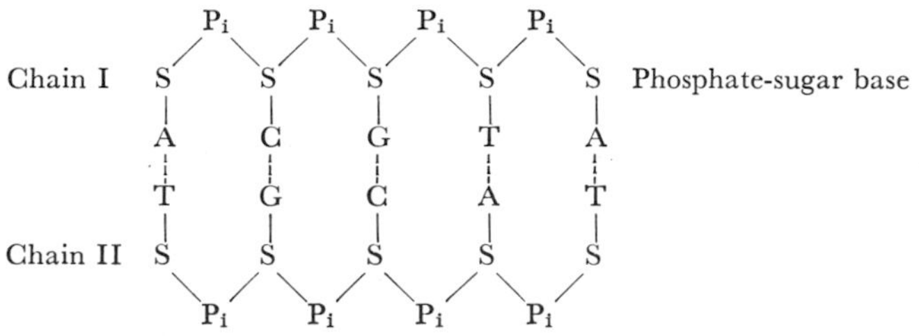

Chain I   S   S   S   S   S   Phosphate-sugar base
P_i P_i P_i P_i

Chain I   A   C   G   T   A
            T   G   C   A   T
Chain II S   S   S   S   S
        P_i P_i P_i P_i

Figure 7-13. DNA structure of cell nucleus of a hypothetical cell before division.

replication was elucidated. The theory was formulated by Watson and Crick and verified by them and a number of other investigators.

DNA consists of a double helix in which the bases are turned inward and held together by hydrogen bonds (see Fig. 7-13). It may be recalled now that for steric reasons the base pairing is always fixed. That means that adenine is always paired with thymine and cytosine always with guanine. During nuclear division the two strands of DNA separate (Fig. 7-14). The dividing cellular protoplasm encloses one strand of DNA before the separation and detachment of the daughter cell are finished.

The single DNA strand alone has to carry all the necessary genetic information to the daughter cell. Development of the daughter cell continues with replication of the DNA strand. The role of DNA polymerase in DNA biosynthesis will be discussed in Chapter 10. At this point we would like to mention only that this enzyme is able to copy the DNA template, which is needed for its activity. The DNA strand obtained from the parent cell may serve as the template, and the DNA polymerase, using the deoxy-nucleoside triphosphates of the daughter cell, synthesizes a DNA molecule the base sequence of which is complementary to the DNA strand of the mother cell.

It is clear that the base composition and sequence of the mother cell and the daughter cells are exactly the same. The unique base sequence of the newly synthesized DNA strand is strictly determined by the base composition and sequence of the template (Fig. 7-15).

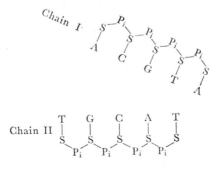

Figure 7-14. Separation of nuclear DNA chains during cell division.

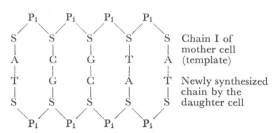

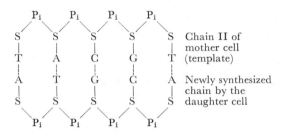

Figure 7-15.   DNA composition of the daughter cells after the division of the mother cell.

Participation of RNA in Protein Biosynthesis

Caspersson discovered that cells which divide rapidly contain large amounts of RNA. This observation was very important because it called the attention to RNA as a participant in the reproductive process. RNA is, however, absent from the sperm and therefore it is most unlikely that RNA plays an important role as a carrier of genetic information in mammalian organisms.

Experimental findings indicate that the physical presence of DNA in certain cells is not required for protein synthesis to proceed for a limited time. During the process of cellular maturation the normoblast, which is a young and immature form of the red blood cell, gradually loses its nuclear DNA. The reticulocyte, which is the next stage in this maturation process, contains no DNA but is rich in RNA and other cellular enzymes. This cell is able to synthesize hemoglobin, the functional protein of the mature red blood cells, in large quantities for days in the complete absence of DNA.

Another line of experiments indicates that the synthesis of cellular proteins does not take place in the nucleus. A major portion of the liver was removed from rats. The remaining liver cells in the animal began to multiply rapidly and in connection with the regeneration process the rate of protein synthesis also increased considerably. For this reason regenerating

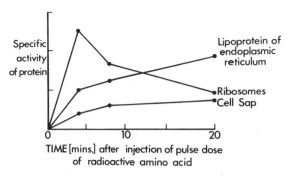

Figure 7-16. The incorporation of amino acids into various cell fractions of the regenerating liver. (From Korner, A.: *In* Mammalian Protein Metabolism. Vol. 1. Edited by Munro, H. N., and Allison, J. B. New York, Academic Press, 1964, p. 185.)

liver is an excellent experimental subject for the study of protein synthesis. Several days after surgery a small amount of radioactive amino acid was given to the rats. Then the animals were sacrificed and the proteins of the various cellular fractions of their liver were isolated. The time sequence study showed conclusively that the radioactivity first appeared in the ribosomal proteins and later in proteins of the cell sap (Fig. 7-16).

Since the ribosomes are in the cellular protoplasm and the "genetic message" is stored in the nuclear DNA, there is an apparent gap in communication between these cellular components. This gap is filled by a special type of RNA with a rapid turnover rate and a variable base composition.

The specificity of these RNA molecules lies in the fact that during their synthesis the genetic message stored in their DNA template is "transcribed" to the RNA molecule. These RNA molecules carry the genetic information to the cellular site of protein synthesis and so earn their name, messenger RNA (mRNA). The mRNA is synthesized by an enzyme called RNA polymerase transcriptase, discovered by Ochoa. This enzyme uses DNA as a template and ribonucleotide triphosphates as substrates. The newly synthesized RNA polymer is a "transcript" of the DNA template which was used for its synthesis. The life span of the mRNA varies from minutes to days depending on the duration of the protein synthesis in a given case. The base pairing is modified because RNA contains uracil instead of thymine. Therefore the adenine group of the DNA template pairs with uracil instead of thymine. Consequently the $\dfrac{A + G}{U + C}$ in the RNA copy must be the same as $\dfrac{A + G}{T + C}$ in the DNA template (Fig. 7-17). This explains how the genetic information stored in the DNA is copied to RNA. Messenger RNA contains 300 to 30,000 nucleotides per molecule. The base ratio varies greatly, just as the base ratio of DNA varies between wide limits. The ribosomes themselves contain RNA independent from the mRNA. This

DNA Template

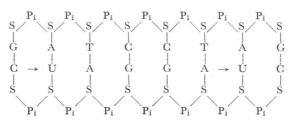

Newly synthesized mRNA

Figure 7-17. Schematic representation of the relationship between the newly synthesized mRNA and its DNA template. The arrows point toward uracils which pair with adenine.

ribosomal RNA has a fairly uniform base composition from species to species, whereas the base composition of the mRNA is widely variable.

Activation of Amino Acids and the Role of Transfer RNA

The exploration of protein synthesis reached a breakthrough when Siekevitz and Zamechnik's group discovered that amino acids can be assembled to protein by a cell-free system. The reaction mixture had the following essential components: activated ribosomes, cell sap, and a complete set of amino acids. Proteins precipitated from the cell sap at pH 5.0 could replace proteins of the whole cell sap in the mixture. They are known as the pH 5.0 enzymes or amino acid-activating enzymes.

An ingenious experiment led to the discovery of another important factor in protein synthesis. Ribosomes and cell sap were incubated with labeled amino acids. After a short period of time one aliquot of this mixture was treated with trichloroacetic acid (TCA). To another aliquot of this mixture a large excess of unlabeled amino acid was added (Fig. 7-18). Since TCA precipitates proteins, upon addition of this compound all enzymatic activity was destroyed. For this reason addition of TCA, as expected, immediately stopped the incorporation of labeled amino acids into ribosomes.

The only difference between unlabeled and labeled amino acids is that the unlabeled compound contains no radioactive atom or atoms. Both unlabeled and labeled amino acids are believed to be used with equal probability for enzymatic reactions. For this reason the addition of a very large excess of cold amino acids to a mixture containing a small amount of labeled amino acids will greatly diminish the probability for the labeled amino acids to participate in a common enzymatic reaction. On the other hand, the addition of unlabeled amino acids does not harm the enzymes so that the incorporation of amino acids into the ribosomal proteins will continue.

If in these experiments all labeled amino acids were free when the mixture was diluted by large excess of unlabeled amino acids, the probability for further incorporation of labeled amino acids in the ribosomal proteins would have been very small because the small amount of labeled amino acids would have to compete with a large excess of unlabeled amino acids for the same enzyme sites. In these experiments, however, after the addition of unlabeled amino acids the incorporation of labeled amino acids into the ribosome continued for a certain period of time; this indicated that the amino acids were not combining with the ribosomes directly, because in that case incorporation would have stopped immediately. Therefore there must be an intermediate between the free amino acid and the ribosome-bound amino acid, and when the unlabeled amino acids were added to the mixture enough labeled amino acid was already present in the intermediate form to continue the incorporation of labeled amino acids into the ribosomes for a period of time. This intermediate has been isolated and shown to be an RNA-amino acid complex. Since this type of RNA transfers the amino acids from the protoplasm to the activated ribosomes it became known as transfer RNA (trRNA).

Transfer RNA is much smaller than mRNA. It usually consists of 80 bases. Studies on the formation of amino acid-trRNA uncovered a number of interesting biochemical events. First of all, the synthesis of this amino acid-trRNA complex is carried out by activating enzymes. We have mentioned them as pH 5.0 enzymes. These enzymes have to recognize the amino acid which will be activated by them. Activation involves the formation of an

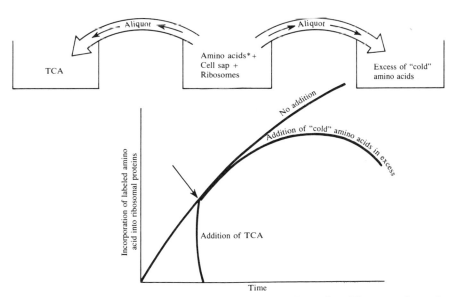

Figure 7-18. Amino acid incorporation into the ribosomal proteins. The arrow shows the time when two aliquots were removed at once from the incubation mixture. One aliquot was added to trichloroacetic acid (TCA) while the second aliquot was mixed with large amounts of "cold" amino acids. Amino acids* = labeled amino acids.

amino acid-AMP complex known as activated amino acid according to the following equation.

A

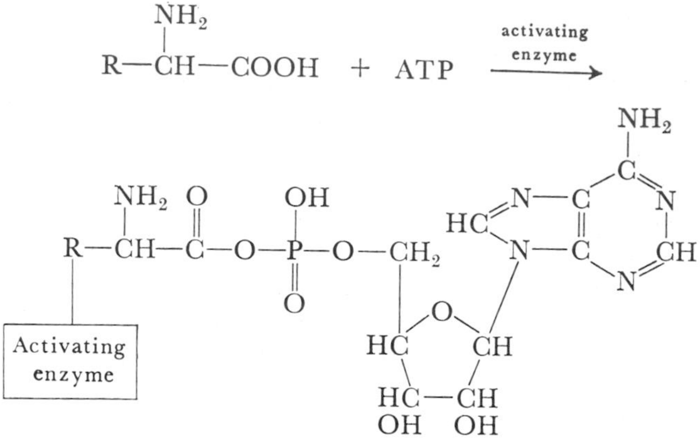

$$R-\underset{\underset{NH_2}{|}}{C}H-COOH \quad + \quad ATP \quad \xrightarrow{\text{activating enzyme}}$$

$$R-\underset{\underset{NH_2}{|}}{C}H-\underset{\underset{O}{\|}}{C}-O-\underset{\underset{O}{\|}}{P}-O-CH_2 \quad \cdots$$

Activating enzyme

Amino acid-adenylate

B Amino acid-adenylate-activating enzyme + trRNA $\longrightarrow$
trRNA-amino acid-adenylate + Activating enzyme

In the first step aminoacyl-adenylate-enzyme complex is formed. In the second step the carboxyl-activated amino acid is linked to the ribose moiety of the last nucleotide on the trRNA. As mentioned before, both steps are catalyzed by the same enzyme. Glycine, alanine, valine, serine, threonine, glutamic acid, aspartic acid, cysteine, methionine, leucine, isoleucine, arginine, lysine, proline, histidine, tyrosine, tryptophan, glutamine, and asparagine are predicted to have separate and special activating enzymes and trRNA's. Therefore the activating enzyme must recognize the specific amino acid as well as the specific trRNA; e.g., tryptophan will combine only with its specific trRNA; this combination is catalyzed by specific tryptophan-activating enzyme.

The trRNA-amino acid complex must find its proper place on the mRNA now attached to the ribosome. There is some experimental evidence that the amino acids so lined up are linked together to form a polypeptide and also removed from the trRNA by one enzyme called polypeptide polymerase (Nathans and Lipmann).

The formation of the native protein from the polypeptide chain or chains is not well understood. One theory is that the polypeptide chain will fold and twist until it arrives at the most stable configuration. This configuration may be determined by the specific amino acid sequence of the polypeptide chain and by the environment of the protein molecule.

THE GENETIC CODE

On grounds of pure logical reasoning the base sequence in a particular DNA was thought to determine the amino acid sequence of the corresponding protein. There are four bases available to code for 20 amino acids. Since a doublet (two bases) code would allow only 16 different combinations, a triplet (three bases) was considered to be the minimal code for a given amino acid. In this case, however, the problem was to see how the possible 64 base combinations among this triplet are used to code 20 amino acids. There are 44 more possible combinations than are theoretically needed. It is possible that some base combinations make sense, and some others do not; for instance, AGC, GCT makes sense but the reverse, CGA, TCG, does not. Only those triplets which make sense are read and used. If this were the case a large number, possibly all 44 unnecessary base combinations, could be thrown out.

It is theoretically possible that more than one code exists for a given amino acid. If this is true, that is, if the code is "degenerate," any number of the 44 unnecessary triplets may have useful function. Some other important questions have also arisen in connection with the reading of the genetic code without ambiguity. In the following base sequence:

ATG CAT GCA TGC ATG C

if one starts at the second letter the same base sequence may be read as:

TGC ATG CAT GCA TGC

and, using the third letter as the beginning, as:

GCA TGC ATG CAT

Needless to say, if such overlaps were possible the coding system would become ambiguous.

Experimental evidence indicates that indeed there are some triplets that make no sense and also that the genetic code is degenerate but that overlaps are not possible. Therefore the code must be read by starting at a fixed point and following the sequence of bases in a given direction. The nature, position, and identity of this fixed point are not known.

At the present time several groups of investigators are involved in research to break the genetic code. One biochemical experimental approach, which earned the Nobel Prize for Nirenberg, will be briefly mentioned. A cell-free system was prepared from *Escherichia coli* which contained the necessary enzymes and ribosomes for protein synthesis. Various synthetic polyribonucleotides were prepared and used as templates for protein synthesis. Under the conditions of these experiments the enzymes transcribed the nucleotide sequence of the synthetic polyribonucleotide template into the newly synthesized polypeptide chain. It has been established that the frequency of UUU triplets in the template was the same as the frequency of incorporation of phenylalanine into the newly synthesized polypeptide chain. Similarly there was a close correlation between the occurrence of GAA triplets in the template and the incorporation of glutamic acid in the newly formed protein. This analysis led to the identification of the coding triplets for many amino acids and provided us with a clue to the genetic code.

THE CONTROL OF PROTEIN SYNTHESIS

Precise control of the quality and quantity of the synthesized protein in the organism is essential for life. In the absence of proper enzyme composition the cellular metabolism is impaired, and the proper function of the organism is disturbed. Such anomalies of amino acid metabolism will be discussed in connection with the intermediary metabolism of the individual amino acid. At this point, however, it must be emphasized that a wide variety of disorders are known which are caused by the lack of an enzyme. These "inborn errors of metabolism" are due to the inability of the organism to synthesize a specific protein (enzyme). Since protein synthesis is genetically controlled, the connection between deficient or altered gene function and disturbed protein synthesis appears to be well founded.

Another type of altered protein synthesis has been mentioned in connection with the discussion of structurally changed hemoglobins. In that case the synthesis of the polypeptide chain undergoes small but functionally important alterations. The inheritance, that is, genetic involvement, in hemoglobinopathies is established beyond doubt.

Experiments performed on lower animals and plants gave important results which conclusively demonstrate that physical or chemical alterations of the chromosome structure may change the ability of the species to synthesize a given enzyme. In the last 5 years an increasing number of experiments have been performed on mammalian cell cultures and whole mammalian organisms which should eventually enable us to understand how the synthesis of a given enzyme is regulated in higher organizations. As it stands, the fundamental concept of the "operon" theory (Jacob and Monod) of biological regulation was derived from experiments performed on bacteria. It was observed that genes which control the synthesis of enzymes involved in a given metabolic pathway often are clustered together within the chromosome. Other results indicate that the synthesis of these enzymes may be accelerated or hindered simultaneously. The operon has been defined as a group of adjacent genes which in their active state give rise to the formation of a single strand of messenger RNA. The genes included in an operon may be subdivided functionally into two classes. The structural genes control the primary structure of the proteins being synthesized, while the operator gene controls the transcription process. An operon has several structural genes and one operator gene. The activity of the operator gene is assumed to be under the control of a regulator gene which need not be in immediate physical contact with the operon. Let us consider the relationship of an operon to cellular metabolism. Assuming that the operon under consideration is responsible for the production of an enzyme (E_1) participating in a metabolic pathway, S_1 is the substrate of E_1 and it is transformed by E_1 to P_1. Both S_1 and P_1 may interact with the operon and thereby influence the synthesis of E_1. S_1 may act as an "inducer," that is, it may

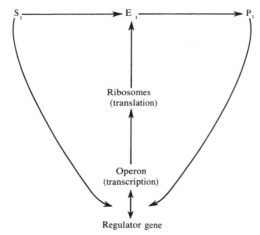

Figure 7-19. Regulation of protein synthesis. E_1 = Enzyme. S_1 = Substrate of E_1. P_1 = Product of E_1.

promote the synthesis of E_1, while P_1 may act in the opposite direction, that is, as a "repressor" (Fig. 7-19).

Both S_1 and P_1 may alter the functional relationship between the regulator gene and the operator gene. Assuming that the steady-state situation involves a certain amount of repression toward the operon, the inducer (S_1) acting on the regulator gene may decrease the repressive function of this gene (derepression) and induce the synthesis of E_1. On the other hand, P_1 may produce the opposite effect on the regulator gene and increase the repression toward the synthesis of E_1. According to this theory, protein synthesis is regulated on the transcription level, since the operon is the unit of genetic transcription. Not only do substrates and products of a metabolic pathway act as inducers or repressors but ever increasing evidence indicates that hormones may have such actions either alone or in combination with the substrate or product of a given enzyme. According to Tomkin, the original operon concept may have to be extended and probably to some extent modified to be applicable for higher organisms. He brought forward results with mammalian cells which indicate that the control of the synthesis of a protein may occur on the level of translation as well as on the level of transcription. This extremely interesting area of the molecular biology is under rapid development.

SUMMARY

The genetic information necessary for the synthesis for a specific protein is stored in the nuclear DNA in mammalian organisms. If the synthesis of this protein is called for, the RNA polymerase "transcribes" the pertinent

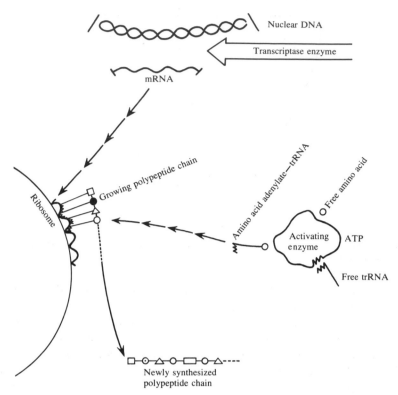

Figure 7-20.   Simplified illustration of some important components of protein synthesis.

genetic message into a newly synthesized RNA molecule. This messenger RNA leaves its nuclear template and moves out into the cytoplasm where it will be attached to the ribosomes. In the cellular cytoplasm the amino acid-activating enzymes synthesize amino acid-RNA intermediates from free amino acids and the so-called transfer RNA molecules. Transfer RNA is able to "read" the code imprinted in the messenger RNA and it "translates" this code into amino acid sequence. The newly synthesized polypeptide chain is detached from the ribosome (Fig. 7-20).

REFERENCES

1. Chargaff, E., and Davidson, J. N. (eds.): The Nucleic Acids. 3 vols. New York, Academic Press, 1955 and 1960.
2. Korner, A.: Protein biosynthesis in mammalian tissue. In Munro, H. N., and Allison, J. B. (eds.): Mammalian Protein Metabolism. Vol. 1. New York, Academic Press, 1964, pp. 177–242.
3. Martin, R. B.: Introduction to Biophysical Chemistry. New York, McGraw-Hill Book Co., 1964.
4. Peacocke, A. R., and Drysdale, R. B.: The Molecular Basis of Heredity. New York, Plenum Publishing Corp., 1967.

5. Perutz, M. F.: Proteins and Nucleic Acids. New York, American Elsevier Publishing Co., 1962.
6. Rich, A., and Davidson, N. (eds.): Structural Chemistry and Molecular Biology. San Francisco, W. H. Freeman and Co., 1968.
7. Tanford, C.: Physical Chemistry of Macromolecules. New York, John Wiley & Sons, 1966.
8. Taylor, J. H.: Selected Papers on Molecular Genetics (A Collection of Reprints with Introductory Material). New York, Academic Press, 1965.

8

SOME ASPECTS OF THE PHYSIOLOGICAL FUNCTIONS OF PROTEINS

The physiological role of proteins in the mammalian organism is manifold. In Chapter 6 a brief summary was given of the general mechanism of enzymatic activity, which may be considered one fundamental biochemical function of proteins. While functioning as enzymes the proteins bind reversibly small organic molecules (substrates and coenzymes) and various ions to some specific protein site (active site). Because of the very nature of the enzymatic functions some of these enzyme-bound small molecules (substrates) are released in a chemically altered form (product) from the enzyme molecule.

The ability of proteins to combine reversibly with small-molecular-weight substances or other macromolecules is also the basis of other physiological functions of proteins, such as transport of molecules, regulation of distribution of ions, and defense against toxic substances. In these capacities, however, the protein does not function as an enzyme and the protein-bound substance need not undergo any chemical modification while combined with the protein.

Physiological functions of proteins that are more complex and not so well understood include coagulation of blood and muscular contraction. During these complicated processes the participating proteins may be involved in enzymatic and nonenzymatic interactions.

Another physiological function of some proteins is their hormonal action. Certain structural characteristics of several protein hormones have been shown to be important in hormone action, but the molecular mechanisms of the protein hormone actions on the target organs are not known.

The transport and regulatory function of proteins with respect to the distribution of ions in the mammalian organism will be illustrated by the description of two proteins, transferrin and the ceruloplasmin. Haptoglobins will also be discussed in this section.

Transferrin

PHYSICAL PROPERTIES. The molecular weight of transferrin is 90,000. It is a glycoprotein with a carbohydrate content of 5.5 per cent (hexose, hexosamine, and sialic acid). Transferrin migrates as a β globulin during electrophoresis. According to the results of physicochemical measurements, it is a rather homogeneous protein.

IRON BINDING. Iron is bound to transferrin by ionic bonds. Two iron atoms are bound per mole of transferrin and 2 moles of bicarbonate ions are taken up into the iron-transferrin complex. This complex is fairly stable in the neutral and alkaline pH range, but it dissociates to its components around pH 4.0. After combining with iron, the transferrin solution, previously colorless, becomes pink. Although the absolute three-dimensional conformation of the iron combining site is not elucidated, there is evidence that its conformation changes when iron is bound to the transferrin molecule. Transferrin obtained from various human sources shows differences in migration in starch gel; 14 different types of human transferrin have been separated by this method. At the present time no clear-cut differences in the amino acid composition of the various transferrin molecules have been established which could account for the differences in electrophoretic mobility.

PHYSIOLOGICAL ROLE. The concentration of transferrin in human plasma is about 240 to 280 μg. per 100 ml. and the iron-binding capacity of 100 ml. plasma is 300 to 360 μg. of iron. Under normal conditions the iron concentration in human plasma is 120 μg. per 100 ml. and almost the total amount is bound to transferrin. The plasma transferrin is therefore only one third saturated with iron. If the free iron concentration of the plasma increases, either by iron absorption from the intestinal tract or by the breakdown of hemoglobin or some other protein molecule, transferrin will combine with the iron and the iron saturation of the transferrin molecule will increase. Conversely, if the free iron concentration of the plasma decreases, iron atoms will dissociate from the transferrin-iron complex. The undersaturated transferrin-iron complex will combine with iron atoms in some iron-rich tissue such as bone marrow or intestinal mucosa and restore its original iron saturation. In this way the transferrin molecule transports iron from one tissue to another and regulates the free iron concentration in the plasma.

Ceruloplasmin

PHYSICAL PROPERTIES. Ceruloplasmin, the copper-binding protein of the blood plasma, has oxidase activity (polyphenol oxidase, monoamine oxidase, and so forth). It is a glycoprotein which migrates electrophoretically as an α_2 globulin. The molecular weight of ceruloplasmin is 150,000, and one molecule of this protein may combine firmly with eight copper atoms. The nature of the copper binding is not well understood.

PHYSIOLOGY AND PATHOLOGY. About 95 per cent of the copper in the blood plasma is bound to ceruloplasmin. The concentration of this protein in the plasma is about 34 mg. per 100 ml. In a congenital disease known as Wilson's disease the synthesis of ceruloplasmin is defective. The plasma ceruloplasmin concentration and copper content are low. The presence of increased deposition of copper in the liver, brain, and other tissues during this disease is well documented, and is attributed to the absence of a sufficient amount of ceruloplasmin to regulate copper transport.

Haptoglobins

Haptoglobins are protein constituents of the normal blood plasma which are able to form rather stable complexes with hemoglobin.

PHYSICAL PROPERTIES. From migration studies in starch gel electrophoresis three major haptoglobin types (Hp 1-1, Hp 2-1, and Hp 2-2) have been discovered. Hp 1-1 migrates as a single band but Hp 2-1 and Hp 2-2 are broken up into several bands during starch gel electrophoresis. The general appearance of these patterns does not change when the haptoglobin is combined with hemoglobin, except for a decrease in the rate of migration of the individual components.

The molecular weight of the Hp 1-1 is between 85,000 and 100,000. Hp 2-1 and Hp 2-2 are believed to be present in polymeric form with molecular weights of 200,000 and 400,000, respectively. The haptoglobins consist of two polypeptide chains. One chain (β) appears to be common in all types while the other chain (α) shows small variations in amino acid composition in the different haptoglobin types. The individual haptoglobin types are under genetic control.

PHYSIOLOGICAL ROLE. The physiological importance of the haptoglobins is their ability to bind hemoglobin. The hemoglobin-haptoglobin complex cannot pass the glomerular filter of the kidney as easily as hemoglobin alone and therefore the complexed hemoglobin is not excreted from the body. Larger amounts of excreted hemoglobin may damage the tubular system of the kidneys and the haptoglobins may prevent such complications. The concentration of haptoglobin in normal persons is between 90 and 120 mg. per 100 ml. The absence of haptoglobin from the plasma of normal persons varies from 1 to 2 per cent of the population in European countries to 32 per cent in some parts of the African continent.

SOME CONCEPTS OF IMMUNOCHEMISTRY; ANTIGEN-ANTIBODY INTERACTION

The mammalian organism is able to produce specific proteins to neutralize certain foreign proteins and other agents that invade the tissues of its body. Any substance which may elicit the formation of such specific neutralizing proteins is called antigen. The specific antigen-neutralizing proteins which are produced by the antigen-stimulated organism are called antibodies. The antigen-antibody complex may have a biological effect different from the one which would have been elicited by the antigen alone. For example, the toxin of *Corynebacterium diphtheriae* has a harmful effect on the mammalian organism. However, when the toxin is combined with a specific antitoxin (this is an antigen-antibody interaction), it has no harmful biological effect at all (it has been neutralized). On the other hand, bovine serum albumin injected intravenously into another species is quite harmless provided that no antibodies against it are present in the blood or other tissues of the recipient; if there are antibodies, the antigen-antibody interaction may have harmful effects, and indeed may even kill the recipient animal. The neutralization or interaction of antigen with antibody may be considered generally as a protective mechanism guarding the organism against infections. Like many other protective mechanisms in the body, it may be "misprogrammed" so that the biological consequence of its operation is disease instead of protection from disease.

If the antigenic substance is a part of a subcellular structure or a cell, the antigen-antibody interaction may lead to a clumping of these antigen-containing structures. This phenomenon is called agglutination. Antigen-antibody interaction on the surface of some cells may result in the lysis of the cells, but this interaction requires the participation of a third component, complement, which is normally present in the blood.

Antigens; Antigenicity

The chemical structure of the antigens is highly diverse. Proteins are the oldest known and most extensively studied antigens. The potency of an antigen to elicit the production of antibodies in the host organism is called antigenicity. In general, the concept of "strong" antigen versus "weak" antigen is relative. Any given protein (or antigen) may be antigenic in one species and nonantigenic in another species. Furthermore, most antigens cause more intensive antibody formation in some animals and less antibody production in others. The responsiveness of the host is of paramount importance in deciding the antigenicity of a given compound. In spite of this, considerable effort has been directed toward characterizing the chemical qualities which are important in determining whether or not a substance may act as an antigen.

First of all, it has been established that the size of the antigen is important with regard to its antigenicity. Apparently there is an ill defined minimum size below which antigenicity vanishes. Oxytocin and vasopressin, with molecular weights of about 1000, are nonantigenic. Insulin, molecular weight 24,000, is a weak antigen, while serum albumin and most of the serum globulins (M.W. 60,000 to 150,000) are good antigens. On the other hand, hemoglobin, which has a molecular weight and shape not very different from those of serum albumin, is a weak antigen. Collagen, which has a higher molecular weight than the previously mentioned proteins, is also only very weakly antigenic. The size of the molecule is important but is by no means sufficient alone to determine antigenicity.

It has been observed that in many instances a single chemically homogeneous antigenic protein stimulates the production of several different antibodies. All of these antibodies interact with the antigen. If the antigen is hydrolyzed with proper proteolytic enzymes it becomes obvious that some of the antibodies interact with a given fragment of the antigen while other antibodies precipitate a different fragment of the same antigen. These results prove that antibodies are not directed against the whole surface of the antigen but rather against certain chemical groups on the surface of the antigenic protein, and thus the concept of "antigenic determinant" emerged. A single antigen may have several determinant groups; the antigenicity of these determinants may be different, and therefore the extent of antibody production elicited by the various determinants is also different.

If the structure of a protein is altered by heat or chemical agents the antigenicity of the various determinants may or may not change, depending on whether or not the steric configuration of these sites is also altered. Some proteins or conjugated proteins, often of bacterial origin, possess harmful effects on the mammalian organism. They are called toxins. Many of these toxins also have antigenic properties. The toxic and antigenic properties of these macromolecules need not be connected with the same chemical groups. Therefore it is possible that chemical alterations of these macromolecules will affect their toxicity without fundamentally changing their antigenicity.

The toxin molecule may be toxic and antigenic; the toxoid molecule, which is a partially denatured form of the toxin, is also antigenic but not toxic (Fig. 8-1). The injection of toxoid to a recipient will lead to the production of

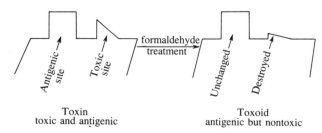

Toxin
toxic and antigenic

Toxoid
antigenic but nontoxic

Figure 8-1. Transformation of toxin to toxoid.

specific antibodies without the harmful effects of the parent toxin. The antibodies produced by the injection of a given toxoid interact with the toxin (from which the toxoid was prepared) and protect the organism from the harmful effects of the toxin.

The Chemical Nature of Some Antigenic Determinants

The use of modified proteins as antigens produced important results regarding the nature of the antigenic site and antigen-antibody interaction. Diazotized sulfanilic acid or similar diazo reagents form diazo compounds primarily with the tyrosine and histidine residues of proteins. Such diazo proteins may be used as antigens. The antibody produced with these antigens shows a high degree of specificity toward the diazo compounds but not toward the protein which has been diazotized. Landsteiner showed that diazotized horse protein injected into a rabbit produced antibodies that would precipitate similarly diazotized chicken proteins but that gave no reaction with the original undiazotized horse proteins. Such experiments with modified protein antigens indicate that antibodies combine with some specific chemical groups on the antigen. Diazotized sulfanilic acid alone does not stimulate antibody production, but combined with protein it becomes the antigenic determinant. There are many small-molecular-weight compounds which are nonantigenic unless they are bound to a macromolecule (usually protein). These compounds are called haptens. The relative importance of certain protein side chains with respect to the antigenicity has become somewhat clearer as the result of experiments performed with polyamino acids. Linear homopolymers of various amino acids were found to be nonantigenic. It was found, however, that the addition of 2 per cent tyrosine to a multichain of combined polyalanine and polyglycine transformed the antigenicity of this heteropolymer from weak to strong. This experiment clearly implicates the tyrosine residues as important factors in the antigenicity.

It was discovered that capsular carbohydrates of pneumococci are precipitated by antisera. These carbohydrates alone are nonantigenic, but coupled with capsular proteins, the carbohydrates become the antigenic determinants. The capsular polysaccharides of various antigenically different pneumococci are also different chemically. The antigenic specificity of the capsular polysaccharides plays an important role in the diagnostic determination of these bacteria.

Severe, occasionally fatal, complications of blood transfusion demonstrated that incompatibility may exist between the bloods of two individuals. Indeed it has been established that the serum of some recipient blood may agglutinate and hemolyze the red cells of the donor's blood. Landsteiner discovered that the red cells contain antigenic substances. These antigenic substances or blood group substances were called type A and type B agglutinogens. Red blood cells contain A, B, A + B, or no agglutinogen. The

Table 8-1. *Red Blood Cell Types*

Red blood cell type (agglutinogen)	A	B	AB	O
Antibodies in the serum (agglutinin)	b	a	o	ab

blood type is called accordingly A, B, AB, or O. The serum of an individual with type A red blood cells contains anti-B antibodies and that of a person with type B red cells has anti-A antibodies. The serum of a person with O blood has both and the serum of a person with type AB has none of these antibodies. It is evident that blood transfusion is safest if it avoids interaction of an agglutinogen with the matching agglutinin (e.g., A with a, or B with b). The substances in the red blood cells which are responsible for their antigeneity are glycolipids, and the carbohydrate portion is the antigenic determinant. Chemically, blood group substances A and B are closely related. Their terminal oligosaccharide contains galactose, N-acetylglucosamine, N-acetylgalactosamine, and fucose residues. With the proper enzymes, A may be converted to B and vice versa. Acetylation of substance B yields substance A. These small differences in the chemical composition of the blood group substances are very important biologically. There are several other blood group systems known; all are based on the presence or absence of certain additional antigenic substances in the red blood cell (e.g., Rh, M, N). The chemical composition of these additional blood group substances is not well characterized.

Nucleic acid may become antigenic but not to the extent that proteins do and not as often. It is interesting to note that the sera of patients with a grave and obscure disease called lupus erythematosus precipitate DNA of various origins. Lipids, fatty acids, and triglycerides are not antigenic by themselves, but they may become a part of an antigen if they are combined with proteins or glycoproteins.

Heterogeneity of Antibodies

Electrophoretic studies using various supporting media have shown heterogeneity in the population of antibodies produced by a single antigen. Heterogeneity is caused by the various antigenic determinants on the antigen which elicited the production of the antibody. Another factor believed to be responsible for the heterogeneity of antibodies is the multicellular nature of antibody production.

Electrophoretic and ultracentrifuge studies of various immune sera showed that there are three classes of antibodies present in most sera: the 7S γ_2 globulins, with a molecular weight of 150,000, the 9 to 11S γ_1A globulins, with a molecular weight of 300,000, and the 19S or γ_1M macroglobulins whose molecular weight is probably 1,000,000.

Structure of Antibodies

The structure of the 7S γ_2 immunoglobulins has been studied in great detail. Partial digestion of these antibodies by papain in the presence of cysteine gives three fragments (I, II, and III). Fragments I and II have molecular weights of 50,000 and are able to combine with the antigen; fragment III, with a molecular weight of 80,000, does not combine with the antigen (Fig. 8-2). In another study the reduction of five S-S bridges of the immunoglobulin with mercaptoethanol and alkylation of the SH groups by iodoacetamide produced four fragments which were separated by gel filtration. The two smaller peptide units are called light (L) chains, and the two larger, heavy (H) chains (Fig. 8-2). The H chains (M.W. 50,000) retain more of the original antibody activity than the L chains (M.W. 20,000) whose antibody activity is weak. Combination of an H chain with an L chain gives much greater antibody activity than the sum of the antibody activity of the separated L and H chains. These results indicate that perhaps both H chain and L chain contribute to the antigen combining site. Since two H and two L chains were isolated per antibody molecule the bivalence of the antibody was further substantiated.

This experimental approach established the approximate structure of the 7S γ_2 antibodies. Figure 8-2 shows the relationship of the various subunits and the presence of two equivalent antigen combining sites. Electrophoretic and amino acid sequence studies revealed that L chains of the "purified"

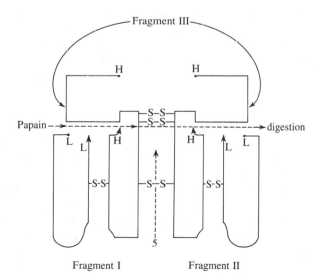

Figure 8-2. Schematic diagram of a 7S γ_2 immunoglobulin. H = Heavy chain; L = light chain; ● = amino terminal of a chain; ↑ = carboxyl terminal of a chain; arrow 5 = a critical disulfide bridge which has to be broken in connection with the papain digestion to obtain fragments I, II, and III.

antibodies are heterogeneous. About 100 amino acids on the C terminal sequence of the L chains of human antibodies show an invariant amino acid sequence, whereas the N terminal sequence is highly variable. Similarly, about 200 amino acids from the C terminal sequence are invariant in the H chain but here too the N terminal sequence shows a great deal of variability. Mild reduction of the 19S antibodies produces 7S antibodies, which retain their immunological activities. The L chains of the 19S macroglobulins appear to be similar to the L chains of the 7S immunoglobulins but the H chains of these two classes of immunoglobulins are distinctly different.

The experimental results described shed some light on the anatomy of antibodies. It has been firmly established that a basic structural similarity exists among antibodies. This similarity involves the number and approximate molecular weight of the amino acid chains and also the invariant C terminal amino acid sequences on homogeneous antibodies. On the other hand, these results also demonstrate that wide variation exists in the amino acid composition of various antibodies (N terminal sequences).

The Formation of Specific Antibodies

In spite of the increasing knowledge about antibody structure and antigenic determinants, a fundamentally biological problem remains unresolved. It is unknown how the antigenic determinant transmits information to the synthetic machinery in order to initiate or speed up the production of specific antibodies.

The proposition that the antigen directs the folding of an already existing polypeptide chain is a possibility but it cannot account for the differences in amino acid composition of the different antibodies. These theories also may have to explain how physical contact between the antigen template and the synthesized antibody is terminated since the antigen-antibody combination is fairly strong.

The "selective" theories imply that the antigen selects among some preformed "informational molecules." In order to satisfy the specificity of the large numbers of antigens, the number of such preformed informational molecules must be enormously large, and it is difficult to reconcile this possibility with the economy of organism.

According to the "instructive" theories, the antigen would, by direct contact, transfer the information to the DNA, RNA, or a polypeptide chain. A great difficulty here is to explain any meaningful communication between the wide diversity of the antigenic determinants and the general code of protein synthesis. According to a modified "selective" theory, the antigen would form a complex with some "preformed peptide chains" or segments of chains in the antigen-producing cell. The complex so formed would stimulate specifically the synthesis of those chains which were taken up into the "antigen-preformed chain" complex. The newly synthesized chains or segments of chains would be assembled like other proteins to yield specific antibody molecules. This very attractive theory needs further experimental verification.

Size and Configuration of the Antigen Combining Site

The size, composition, and configuration of the antigen combining sites of antibodies is not well known. In recent years several excellent methods have been developed to study the structure of the combining sites on antibody molecules. One of these methods is called the "differential labeling" technique. First the antibody is allowed to react with a hapten molecule. In the second step a suitable reagent (R) is added to the antibody-hapten complex. This reagent may form covalent bonds with specific amino acid side chains of the antibody molecule which are not protected by the hapten. The hapten, which is attached to the combining site will prevent any interaction between the reagent and those amino acid side chains that belong to the combining site. After removal of the hapten molecule the radioactive sample of the reagent is permitted to react with the antibody. This radioactive R will interact with those specific amino acid side chains which were previously protected by the hapten, that is, the groups on the combining site. The use of various reagents has permitted identification of the groups involved in side chain structure of the combining sites (Fig. 8-3).

It is hoped that to the rapidly increasing body of information will be added the findings necessary to elucidate the size and structure of the antigen

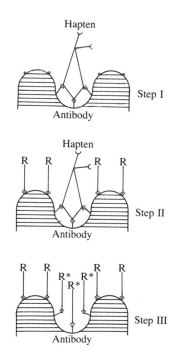

Figure 8-3. Differential labeling.

$\dfrac{R}{|}$ = Cold reagent;

$\dfrac{R^*}{|}$ = radioactive reagent.

combining site. This will enable us to gain further insight into the formation and function of antibodies.

Antigen-Antibody Interaction

If a proper amount of antibody is added to antigen within a suitable pH range (pH 4.0 to 9.0), temperature (0 to 40°C.), and salt concentration (0.05 to 0.25 M) the interaction between antigen and antibody will lead to the formation of precipitate. In the mechanism of this protein-protein interaction the role of several factors must be considered:

1. Complementariness between the antigenic site of the antigen and the combining site of the antibody.

2. Nonspecific electrostatic forces due to the ionized side chains of the proteins.

3. Short range forces, such as dipole induction and London forces.

The antigen-antibody combination may be adequately explained on the basis of the immunological specificity ensured by the mutual complementariness of the combining sites. One important factor in the precipitation reactions is the relative proportion of antigen to antibody in the reaction mixture.

If antigen and antibody are mixed in various proportions and the total amount of precipitate is measured, a curve like that shown in Figure 8-4 is obtained. The zone where precipitate formation is maximal is called the equivalence zone. This is preceded by the zone of antibody excess, in which a part of the total antibody remained in the supernatant. The descending part of the curve represents the zone of relative antigen excess. In this latter zone a part of the antigen-antibody precipitate is redissolved and the supernatant contains free antigen and soluble antigen-antibody complexes. At the present time the exact cause for the solubilization of the antigen-antibody

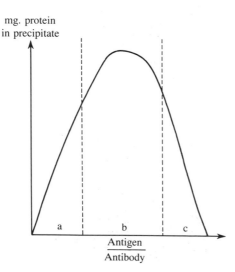

Figure 8-4. Precipitation of antigen-antibody complex as the function of antigen-antibody ratio.

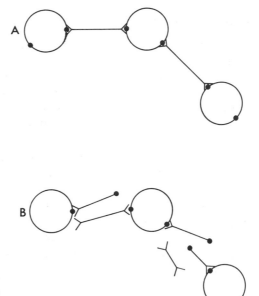

Figure 8-5. Agglutination. *A*, Agglutination by bivalent antibodies. *B*, Inhibition of agglutination due to the presence of monovalent antibodies.

precipitate by excess antigen is not known. There is some evidence that in the equivalence zone the antigen-antibody complex has a tendency to form "branched-chain" polymers and precipitate. In the presence of excess amount of antigen this tendency may be repressed and the combined antigen-antibody molecules remain in solution.

Agglutination

Frequently the antigen is firmly bound to a cellular structure. Such antigens are often called agglutinogens. Antibodies formed against these cell-bound antigens may interact with the agglutinogen and precipitate the cell. The specific antibodies (agglutinins) that are able to cause precipitation have two combining sites (are bivalent) which interact with the specific agglutinogen sites on two different cells. Since the formation of a few links between two cells is sufficient for agglutination, a relatively small amount of bivalent antibody is able to cross-link large numbers of cells and produce agglutination (Fig. 8-5*A*). Monovalent antibodies cannot produce agglutination; rather, they inhibit the agglutinating effect of the bivalent antibodies since they may occupy the same sites which would be suitable for the bivalent antibodies (Fig. 8-5*B*).

The agglutination reaction is widely used in medical laboratories. Determination of blood groups and identification of many bacterial species are examples of the importance of these agglutination reactions.

Complement Fixation

If well washed sheep red blood cells are injected into rabbits, antibodies are formed in the rabbits against the antigenic sites of the sheep erythrocytes. The rabbit immunoserum so produced hemolyzes the sheep erythrocytes in vitro and in vivo. If the rabbit serum is exposed to 56°C. for a short time, hemolysis of the sheep erythrocytes is inhibited. The heat-treated rabbit serum regains its hemolytic activity if small amounts of fresh guinea pig serum are added to the mixtures. Subsequent work shows that the sera of many species contain a rather complex heat-labile system called complement which is often associated with the antigen-antibody complexes. In the mixture just described the rabbit serum contains the specific antibodies and the sheep erythrocytes possess the antigenic sites. In the absence of complement the antigen-antibody combination proceeds (the erythrocytes become "sensitized") but hemolysis is delayed until complement (fresh serum of a suitable species) is fixed to the antigen-antibody complex. Recent investigation shows that complement consists of several different proteins (see Fig. 8-6).

The mechanism of action of complement is obscure. It has to be attached to the sensitized red cell in order to bring about the hemolysis. Some electron microscopic evidence indicates that hemolysis is the result of "single-hit" action of complement. Single holes were found on the surface of hemolyzing red blood cells corresponding to those loci where complement was fixed.

In spite of our ignorance about the precise mechanism of immune hemolysis, "complement fixation" is extremely important in the medical laboratory diagnosis. The Wassermann reaction for the diagnosis of syphilis is based on complement fixation. This reaction discloses the presence of "specific antibodies" produced by the disease which react with a suitable antigen. The performance of the Wassermann test involves the following steps (see also Fig. 8-7). The complement in the patient's serum is inactivated at 56°C. During the first incubation measured amounts of antigen and

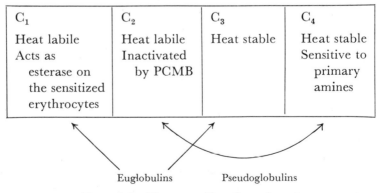

C_1	C_2	C_3	C_4
Heat labile	Heat labile	Heat stable	Heat stable
Acts as esterase on the sensitized erythrocytes	Inactivated by PCMB		Sensitive to primary amines

Euglobulins   Pseudoglobulins

Figure 8-6.   The composition of complement.

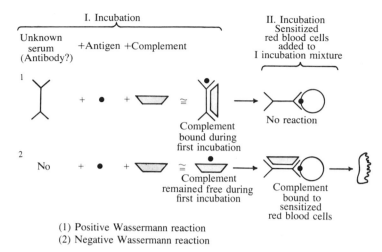

(1) Positive Wassermann reaction
(2) Negative Wassermann reaction

Figure 8-7. Schematic representation of the Wassermann reaction.

complement are added to the patient's inactivated serum.* If the patient's serum contains specific antibodies these will react with the antigen and fix the complement. This reaction does not lead to any visible change in the mixtures. A second incubation of these mixtures with sensitized sheep erythrocytes† discloses whether or not the complement was fixed to the antigen-antibody complex if such a complex has been formed. If complement was fixed during the first incubation the sensitized sheep erythrocytes will not be hemolyzed. This means that the patient's serum contained antibodies characteristic of a syphilitic infection. If such antibodies were not present in the patient's serum, the complement was not fixed during the first incubation and is available to hemolyze the sensitized sheep erythrocytes (Fig. 8-7), and the Wassermann reaction is negative.

THE COAGULATION OF BLOOD

The coagulation of blood is an important defensive mechanism for the mammalian organism. If this mechanism is defective minor cuts and injuries cause serious bleeding and any surgical procedure carries a high risk of major hemorrhage originating from minute blood vessels.

The healthy mammalian organism possesses all components necessary for the coagulation of the blood. At the same time, however, the organism needs means by which spontaneous unnecessary activation of the coagulation process is inhibited. The formation of blood clots within the blood vessels (thrombosis) is a dangerous condition. Clots may obstruct vitally important

---

* Inactivated serum does not contain complement.
† Sensitized sheep cells contain absorbed specific antibodies.

blood vessels, causing severe functional impairment in the body or even the death of the organism.

We know of two ways by which the mammalian organism prevents this serious complication. First, several clotting factors are normally present in an inactive form and have to be activated before they can participate in the clotting process. Second, it has been shown that the plasma contains compounds known as anticoagulants which inhibit various steps of clot formation.

During the coagulation of blood, proteins function as enzymes, substrates, activators, and inhibitors, and they participate in nonenzymatic interactions, such as polymerization, which are also necessary. The clotting mechanism and its regulation are fascinating examples of the physiological function of proteins.

The blood is a two-phase system. It consists of cells and plasma. After coagulation some of the plasma proteins are used up or participate in the insoluble clot. The cellular elements are partly trapped in the clot while the remaining components are found in the soluble phase called serum.

The fundamental process in the coagulation of blood is the enzymatic transformation and spontaneous polymerization of a soluble plasma protein called fibrinogen into insoluble fibrin filaments. The highly specific proteolytic enzyme, thrombin, which is responsible for the initiation of this transformation, is present in blood in an inactive form called prothrombin. Fibrinogen, prothrombin, and thrombin have been isolated and purified and their physical-chemical properties are quite well defined. For this reason the chemistry of the fibrinogen-fibrin transformation is also well understood.

Fibrinogen-Fibrin Transformation

Fibrinogen is a rodlike, water-soluble protein. Its physiological concentration in the plasma is 300 to 400 mg. per 100 ml. The solubility of this protein in salt solutions is low, and this is the basis for its isolation from the plasma. The molecular weight of bovine fibrinogen appears to be about 340,000. Electron microscopic studies indicate that its length is about 400 to 500 Å. Glutamic and aspartic acid accounts for 27 per cent of its total amino acid composition and it is likely that the stability of fibrinogen in solution is insured by the large number of acidic side chains. Fibrinogen contains small amounts of hexose and hexosamine. An intact carbohydrate moiety is a prerequisite for the clotting ability of the fibrinogen molecule.

Prothrombin is available in highly purified form. Its concentration in bovine plasma is about 10 to 15 mg. per 100 ml. The molecular weight of the protein is about 68,000. It is a globular protein having an isoelectric point of 4.2. Prothrombin contains 11.6 per cent carbohydrate, mainly sialic acid, galactose, mannose, and hexosamine. Seegers and his collaborators observed that purified prothrombin, when dissolved in 25 per cent citrate solution, "spontaneously" converted to thrombin. This finding proved conclusively that prothrombin is the sole macromolecular precursor of thrombin. It was also observed that during citrate conversion prothrombin is split up into several products. According to the view of Seegers and his colleagues, three of these split products may participate in the clotting process. Another important concept which emerged from the citrate conversion experiments is that prothrombin is a very labile protein molecule which under suitable conditions may dissociate to some smaller biologically active units. Compounds that promote such dissociation were named procoagulants; those that retard it are called anticoagulants.

Thrombin, unlike its inactive precursor prothrombin, is not normally present in any measurable amount in the normal blood plasma. Physicochemical measurements show that bovine thrombin is a globular protein with a molecular weight of 33,000. This protein is an enzyme which has clotting and esterase activity. The time necessary to produce solid fibrin clot from a standard fibrinogen solution is used to quantitate the clotting activity of thrombin.

In thrombin-catalyzed proteolysis the N terminal glutamic acid residues of fibrinogen are replaced by glycine residues and two small peptides are split off the fibrinogen molecule.

Electrophoretic studies showed that the fibrin monomer (fibrin monomer = fibrinogen less fibrinopeptide) is not so acidic as the parent fibrinogen molecule. It was suggested that fibrin monomers, unlike the fibrinogen molecules, attract each other; such electrostatic attraction would be primarily responsible for the polymerization of these proteins. The exact nature and number of bonds formed during the polymerization process are still uncertain. There is some evidence that the fibrin polymer is primarily stabilized by hydrogen bonds. According to this theory 19 hydrogen bonds are formed per dimer, the donors being lysines and tyrosines, and the acceptors, histidine side chains. The possibility that coordinate covalent bond formation rather than hydrogen bonds are responsible for fibrin formation has also been considered.

Intensive physicochemical studies failed to reveal any major structural differences between fibrin and fibrinogen. The α helical content is about the same (30 per cent), and the peptide maps are identical except that the fibrinopeptides are missing from the fibrin. This result may mean that fibrinogen does not undergo any major molecular alteration or structural rearrangement during the polymerization process.

Fibrin threads formed in vitro from highly purified fibrinogen and thrombin dissolve easily in 30 per cent urea. On the other hand, the fibrin clot produced by coagulating plasma remains insoluble in the same solvent. This observation led to the discovery and isolation of a substance from the serum and other tissues which is able to convert the urea-soluble fibrin to urea-insoluble fibrin. This compound is called "fibrin-stabilizing factor" or "fibrinase." Its molecular weight is thought to be about 350,000. The mechanism of action of the fibrinase was shown to be a transpeptidation or transamination that results in covalent bond formation between amino donor

(A)   (D)

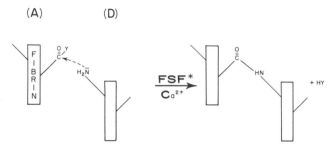

Figure 8-8. Crosslinking of fibrinogen. D = Donor; A = acceptor; FSF = fibrin-stabilizing factor. (Courtesy of L. Lorand.)

groups and carboxyl receptor groups. Donor and receptor groups are located on different fibrin monomer units (see Fig. 8-8). This cross-linking of the fibrin monomers confers a greatly increased stability to the fibrin clot.

The formation of fibrin from fibrinogen may be summed up as follows:

1. $(\text{Fibrinogen})_n \xrightarrow{\text{thrombin}} \left(\dfrac{\text{Fibrin}}{\text{monomer}}\right)_n + (\text{Fibrinopeptide A})_n$

$+\ (\text{Fibrinopeptide B})_n$

2. $(\text{Fibrin-monomer})_n \rightarrow \text{Fibrin}$

3. $\text{Fibrin} \xrightarrow{\text{Fibrinase (Ca}^{++})} \text{Cross-linked fibrin}$

DEFECTS OF FIBRIN FORMATION. If fibrinogen is missing from the plasma or if it is present only in abnormally low concentration the plasma may become incoagulable. About 60 cases of complete absence of fibrinogen from the blood plasma have been reported in the clinical literature. Congenital afibrinogenemia is an inherited disease with a poor prognosis. Several cases are recorded in medical literature of absence of fibrin-stabilizing factor from the plasma. In these cases the coagulation system appeared to work normally but the clot formed was defective. These patients developed hemorrhagic tendencies. The absence of fibrin-stabilizing factor in the human plasma may be congenital or acquired in connection with other diseases.

Prothrombin-Thrombin Conversion

It was mentioned before that under normal conditions thrombin is not present in the blood. The precise mechanism of the biological conversion of prothrombin to thrombin is not known. The participation of several factors, probably distinct plasma proteins, has been inferred from clinical findings. These clinical observations have usually been made in patients afflicted with bleeding tendencies. Laboratory tests performed on the blood of such patients indicated that the prothrombin-thrombin conversion was defective. Conversion became normal in vitro and in vivo when normal plasma was mixed with the blood of the patient. Therefore intensive research has been conducted to isolate and characterize all those components (factors) of the normal plasma or serum which are necessary for the prothrombin-thrombin conversion.

Some of these factors have been purified to a great extent (Factors V* and VIII) but the majority of the proposed factors (Factors VII, IX, X, XI, and XII) have not been completely purified and therefore their biochemical characterization is poor. This fact may simply reflect the difficulty imposed upon the purification of a protein which is present only in trace amounts in a protein-rich tissue. On the other hand, future experiments may reveal that some of these "prothrombin-converting factors" do not exist at all.

EXTRINSIC CLOTTING MECHANISM. It should be noted that the citrate conversion of prothrombin to thrombin takes 12 hours, whereas the so-called "biological activation" is finished in 12 seconds.

According to the classic theory, the prothrombin-thrombin conversion is mediated by a factor which is liberated from damaged tissues and blood cells. First this factor was called thrombokinase but later the name thromboplastin became more popular. A large number of chemically different agents obtained from different sources appear to have potent thromboplastin activity. Brain and lung extracts, milk, and Russell's viper venom are all widely used as thromboplastins. Attempts at chemical characterization of thromboplastins have been frustrating. Usually purification of the various tissue extracts results in a progressive loss of thromboplastin activity. In spite of that it appears to be fairly well established that the active principle is a phospholipid or a phospholipid-protein combination. It is well documented that ionized Ca^{++} is essential for the thromboplastin activity.

To measure the conversion of prothrombin to thrombin in the presence of thromboplastin the following mixture has been used.

Citrated or oxalated plasma + Ca^{++} + Thromboplastin → Clot
$$\text{formation}$$
(The formation of fibrin threads was timed.)

It was either assumed, or measured in a separate test, that the plasma contains sufficient fibrinogen and only the plasma prothrombin concentration is limiting in the presence of sufficient amounts of thromboplastin and Ca^{++}. This test is often called the "one-stage prothrombin time" and, in spite of its obvious shortcomings, is widely used in clinical laboratories. One shortcoming became evident when certain plasmas or prothrombin preparations which gave a weak reaction or no reaction at all with thromboplastin + Ca^{++} reacted readily if "prothrombin-free"† fractions of normal plasma

---

* Fibrinogen = Factor I; thrombin = Factor II; thrombokinase = Factor III; Ca^{++} = Factor IV.

† Some of those so-called plasma subfractions were not always absolutely prothrombin-free; e.g., the plasma fraction containing Factor VII always has some prothrombin contamination.

were also incorporated into the mixtures. It became obvious that under the conditions of this test prothrombin may not be the only limiting factor in fibrin formation because factors other than thromboplastin and Ca^{++} are also needed for the "biological conversion" of prothrombin to thrombin. In the one-stage prothrombin test the plasma serves as the sole source of fibrinogen, prothrombin, and the additional converting factors. The same plasma, however, may also contain compounds inhibitory for thrombin formation or thrombin action. If such inhibitor is also present in the plasma tested, the result will be a prolonged prothrombin time, just as if either fibrinogen, prothrombin, or converting factors were present in reduced amount in the plasma. Evidently, the one-stage prothrombin test in its original form is unable to distinguish between the absence of an essential component for the thrombin formation and the presence of an inhibitory compound in an otherwise complete system.

These difficulties are to some extent eliminated in the so-called "two-stage prothrombin time" test. Here the plasma which is being tested is diluted. Dilution was found to minimize the effect of inhibitors. In the first stage the so-called converting factors, such as Factor V, Factor VII,* thromboplastin, and Ca^{++}, are added to the diluted plasma. Aliquots removed from this mixture at various time intervals are mixed with fibrinogen solutions in a different test tube and the appearance of fibrin threads in this second mixture is timed. The testing of aliquots removed from the first mixture is continued until the shortest clotting time is measured in the second mixture. Apparently the shortest clotting time in the second mixture discloses the presence of the largest possible amount of thrombin formed in the first mixture.

Discrepancies between prothrombin times measured by the one-stage test and those measured by the two-stage test indicated that factors other than Ca^{++} were also involved in the thromboplastin-induced prothrombin-thrombin conversion. Factors V and VII were discovered to be important in accelerating the conversion of prothrombin to thrombin.

Factor V. One of these factors, Factor V, is a labile protein. It is quickly destroyed at 56°C. and disappears from the plasma during storage. Factor V has been purified several hundredfold. In human plasma the Factor V concentration is 1 mg. per 100 ml. There are great differences in the molecular weight values of Factor V purified by different authors. Figures as diverse as 97,400 and 290,000 have been reported. Factor V, unlike prothrombin, is not adsorbed to barium sulfate suspension;† therefore a $BaSO_4$ suspension-treated, defibrinated, fresh plasma is a convenient source of Factor V. Factor V is consumed during clotting and it is not present in sera. The possibility of congenital lack of Factor V cannot be

---

* Prothrombin-converting factors.

† Citrated or oxalated plasma is mixed with $BaSO_4$ suspension. After a certain time this mixture is centrifuged. Some clotting factors will precipitate with the $BaSO_4$ while others remain in the supernatant.

excluded, but acquired Factor V deficiency is more common. It is often connected with liver disease and malignancies.

Factor VII. The chemical characterization of another important accelerator protein, Factor VII, is poor. It is, however, established without a doubt that it is different from Factor V. Factor VII is not consumed during the coagulation process; therefore it is present in sera. Unlike Factor V, it is adsorbed to $BaSO_4$ suspension. Its electrophoretic migration is similar to those of the β globulins. The Seegers school has brought forward some evidence that a split product of the citrate-treated prothrombin, called autoprothrombin C, is able to normalize coagulation deficiencies caused by Factor VII deficiency. In their opinion Factor VII (and also Factors IX and X) as a distinct serum protein does not exist. There is little doubt that Factor VII as a functional unit of the coagulation system is indeed important. The undecided question is whether Factor VII is a split product of prothrombin or a distinct serum protein. Clinical deficiency of Factor VII is well documented. Addition of Factor V does not correct the deficiency.

There is no positive evidence about the nature of the reaction, or, more probably, reactions, which occur in the mixture containing prothrombin, tissue thromboplastin, Ca^{++}, Factor V, and Factor VII. Probably several interactions take place simultaneously or in close succession between pairs of molecules and between the products of such interactions. The final result is an extremely potent prothrombin-converting enzyme, "prothrombinase." The prothrombin-thrombin conversion with tissue thromboplastin may be represented by the following equation:

$$\text{Thromboplastin} + Ca^{++} + \text{Factor V} + \text{Factor VII} \rightarrow \text{Prothrombinase}$$

$$\text{Prothrombin} \longrightarrow \text{Thrombin}$$

INTRINSIC CLOTTING MECHANISM. It is an old observation that blood drawn into a test tube will clot in 5 to 10 minutes without the addition of tissue thromboplastin. Since the clotting time with added tissue thromboplastin is much shorter (10 to 15 seconds) it was originally thought that a weak tissue thromboplastin may be present as a contaminant in the "intrinsic" system. A time sequence analysis of the "thrombin generation," however, indicated a different mechanism (Fig. 8-9). If a weak tissue thromboplastin is active, one would expect the appearance of a small amount of thrombin right after the blood is shed, with a gradual increase thereafter (see the dotted line in the figure). Experiments proved, however, that after a lag period which lasts for several minutes the thrombin production starts abruptly and increases in an explosive manner. These results are in line with the assumption that the lag period is needed for the activation of a powerful blood thromboplastin system (intrinsic system) which then appears as a functional entity and leads to the rapid production of thrombin.

The following system consisting of an incubation mixture and a testing

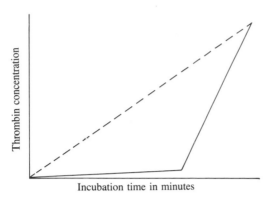

Figure 8-9. Activation of intrinsic thromboplastin.

mixture was found to be useful to study the generation of intrinsic thrombo-plastin activity. The incubation mixture (mixture I) may contain various plasma, serum, and thrombocyte fractions other than prothrombin and fibrino-gen. These latter two components were used in a testing mixture (mixture II) separate from the incubation mixture. The incubation mixture is used to generate a prothrombin-converting principle, intrinsic thromboplastin. The testing mixture serves for the detection and quantitation of the intrinsic thromboplastin. From time to time aliquots removed from the incubation mixture are added to the samples of testing mixture (Fig. 8-10). Thrombin is not generated in the incubation mixture since in the absence of prothrom-bin this is not possible. In the testing mixture, however, thrombin generation may proceed, depending upon the appearance and concentration of a pro-thrombin-converting principle, intrinsic thromboplastin. If the composition of the incubation mixture is varied, a number of "factors" can be "detected" which are believed to play a role in the intrinsic thromboplastin generation. These factors will now be briefly considered.

Factor VIII. The concentration of this factor in human blood is about 1 mg. per 100 ml. Because of the great clinical importance of Factor VIII, intensive research was carried out in many laboratories to purify it. This effort was hampered by the factor's extreme lability and also by the similar precipitation properties of Factor VIII and fibrinogen. Most purified Factor VIII preparations may be contaminated by small amounts of fibrino-gen. The molecular weight of Factor VIII was found to be between 180,000 and 200,000. Electrophoretically it migrates as a β globulin. The concen-tration of Factor VIII rapidly diminishes in stored plasma and it is absent in serum. Factor VIII is not adsorbed to $BaSO_4$ gel. It contains about 3.6 per cent carbohydrates. The absence of this factor is connected with a bleeding defect known as hemophilia. In this sex-linked, recessively in-herited disorder the female serves as carrier but only the male is affected.

Factor IX. Normal plasma probably contains only the inactive precursor of Factor IX. In spite of the fact that 700 fold purification has been reported, the "purified" preparation is still contaminated with prothrombin and

Factor X. Factor IX is adsorbed to $BaSO_4$ gels. The absence of this factor from the blood results in a hemophilia-like disease in both males and females. Purified Factor VIII cannot correct the coagulation defect caused by the deficiency of Factor IX, but the plasma of a hemophilia patient with Factor VIII deficiency may correct the coagulation defect. Conversely, a Factor IX-deficient plasma may reverse the coagulation deficiency caused by the absence of Factor VIII.

Factor X. This factor is present in plasma and serum. It is adsorbed from plasma to $BaSO_4$ along with prothrombin, Factor VIII, and Factor IX. The various preparations of Factor X are usually contaminated with small amounts of Factor VII and Factor IX. Electrophoretically it migrates as an α globulin. Congenital deficiency of Factor X has been reported in both sexes.

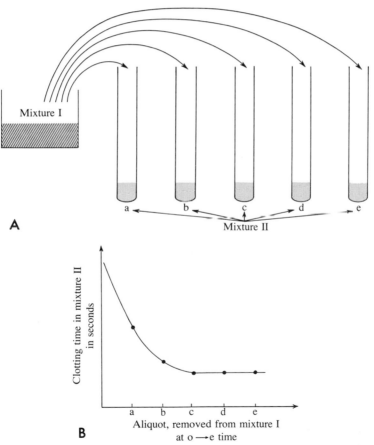

Figure 8-10. Schematic representation of the intrinsic thromboplastin generation test. *A, Mixture I:* Thrombocyte fractions and plasma or serum fractions, or both (no prothrombin, no fibrinogen!). *Mixture II:* Pure prothrombin + pure fibrinogen. *B,* Graphic representation of the results. According to the curve maximum intrinsic thromboplastin was formed in mixture I at C time.

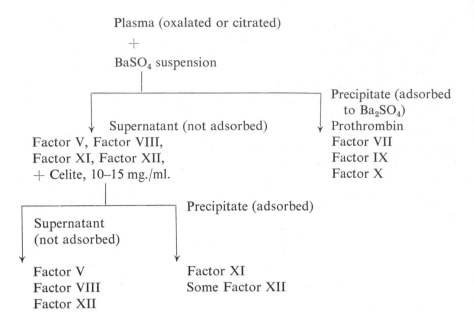

Figure 8-11. An outline of the separation of the various clotting factors using BaSO₄ and Celite suspensions.

Factor XI. Factor XI may be prepared from plasma and serum. It is not adsorbed to BaSO₄ gels and therefore is easy to separate from Factors VII, IX, and X and prothrombin. The chemical separation of Factor XI from Factors I to X is quite successful, but it is a much more difficult problem to separate Factor XI from Factor XII. Factor XI activity is found in fractions which migrate in the β and α regions during electrophoresis. Its absence from the plasma may be responsible for mild to moderate bleeding episodes in both sexes.

Factor XII. This factor has not yet been isolated in the form of a homogeneous fraction and therefore the physicochemical characterization of Factor XII is poor. Congenital deficiency of Factor XII has been called Hageman trait. In spite of the apparent importance of Factor XII in the coagulation process in vitro, Hageman trait is usually not associated with bleeding tendencies.

The use of BaSO₄ suspension in the separation of various clotting factors is schematically illustrated in Figure 8-11.

Platelet participation in blood thromboplastin generation appears to be firmly established. Platelets are fragments of the protoplasm of a large cell called a megakaryocyte and as such they contain large number of enzymes and other biologically active compounds. Several fractions were isolated from the platelets which exerted influence to the clotting process. Platelet Factor 3 appears to be important in blood thromboplastin generation.

Platelet Factor 3. This lipoprotein has been isolated from the cytoplasmic granule of the platelets. Among the phospholipids isolated from this fraction phosphatidylethanolamine and phosphatidylserine are active in promoting plasma thromboplastin formation. The free amino group of the serine and of the ethanolamine and the unsaturation of the fatty acid moiety are essential for the function of Platelet Factor 3 in the clotting process.

At the present time the precise mechanism of intrinsic thromboplastin formation is not known. Some of the theories which were put forward by the leading investigators of this field are important and must be discussed briefly. It seems to be firmly established that the surface of the container in which blood is kept has an influence on the speed of blood clotting. Plasma essentially free from cellular elements will clot in minutes if kept in a glass tube but it will remain liquid for hours if the internal surface of the glass tube is coated with silicon or paraffin. Among the various clotting factors tested Factor XII appears to be activated by such surface action. The chemical nature of this activation is unknown. One proposed scheme for the formation of intrinsic thromboplastin (Davie and Ratnoff) is shown in Figure 8-12.

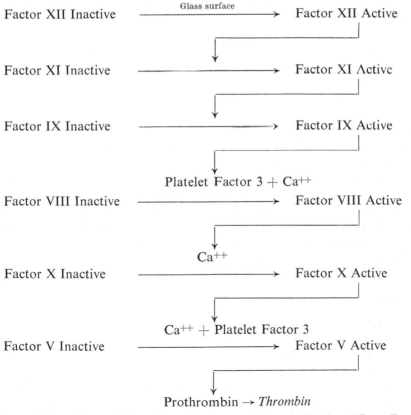

Figure 8-12. Schematic illustration of the "cascade" theory of Davie and Ratnoff.

This scheme, which is often called the "cascade" type of reaction sequences, may account very well for the multiplicity of clinical findings which actually started the search for these various different "factors." One has to recognize the serious uncertainties involved since most of the factors which are believed to participate in the intrinsic thromboplastin formation are poorly character- ized chemical entities. There is very little information regarding any chem- ical difference between the so-called "inactive" and "active" form of the various factors. Needless to say, it is impossible at this time to evaluate or even seriously speculate about the nature of reactions depicted by the various arrows.

According to the theory of the Seegers school, the prothrombin-thrombin conversion, with or without the participation of tissue thromboplastin, is an autocatalytic process. This means that a small-scale prothrombin break- down liberates products that will speed up the degradation process itself (autocatalysis). The extrinsic and intrinsic prothrombin-thrombin conver- sion differs only in the number and nature of the procoagulants which initiate the breakdown of a small amount of prothrombin. Once a certain small amount of prothrombin has broken down and thereby a certain amount of split product has been set free, the degradation process will catalyze itself (see Fig. 8-13).

Figure 8-13 shows an oversimplified version of this autocatalytic process. Under suitable environmental conditions small amounts of prothrombin break down to thrombin and other active split products of the parent pro- thrombin molecule, and these products speed up the conversion of prothrom- bin to thrombin.

It was pointed out in the early part of this section that the Seegers theory provided new concepts in the explanation of the blood clotting process. One of these concepts is that several supposed plasma factors are in fact breakdown products of the prothrombin molecule. There are some indica- tions that this theory has certain difficulties in common with the "cascade" theory. Some of the prothrombin split products are not well characterized chemically. The nature of interaction between the procoagulants and pro- thrombin, as well as between the various split products of prothrombin and the prothrombin molecule, is not clearly understood. Further research will without doubt clarify these crucial points.

Anticoagulants

It has been shown in the previous section that the plasma contains all the components necessary for its conversion from liquid to solid (clotting factors). At the same time plasma has a well documented ability to inhibit the clotting process. Unfortunately the chemical characterization of these anticoagulant substances is poor, and our knowledge about their mechanism of action is also incomplete. In the following paragraphs we will discuss several naturally occurring and some artificially synthesized anticoagulants.

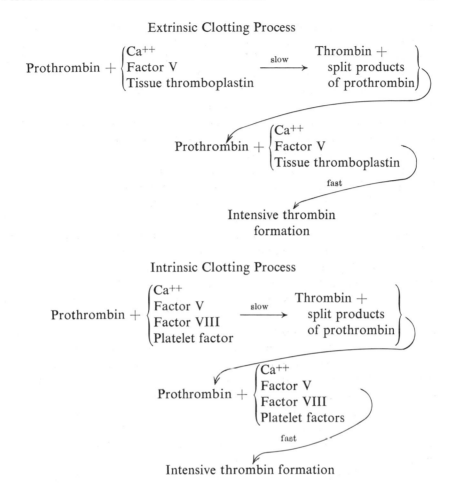

Figure 8-13. Simplified scheme of extrinsic and intrinsic coagulation mechanisms according to the concept of the Seegers school.

As in the case of the "discovery" of many clotting factors, clinical observations contributed a great deal to the recognition of anticoagulant activity on blood. This anticoagulant activity is explained by some authors on the basis of circulating anticoagulants, the action of which may be directed preferentially against some clotting factor. Anticoagulant action against Factors X, IX, VII, and others has been described. The circulating anticoagulants in these cases are poorly characterized. Some workers believe that they act like antibodies, neutralizing the individual clotting factors; others prefer an enzymatic mechanism of action as the explanation.

It was shown that a plasma glyceride, called "inhibitor source material" inactivates Factor VIII. Thrombin and Ca^{++} facilitate the interaction between this "inhibitor source material" and Factor VIII. Some workers believe that "inhibitor source material" has great physiological importance.

Under normal conditions blood plasma contains most of the factors which are necessary for clotting. If sufficient amounts of Platelet Factor 3 escape from the platelets, in the presence of Factor VIII, Factor V, and prothrombin, thrombin formation may proceed. It was proposed that "inhibitor source material" may control the interaction of Platelet Factor 3 with Factor VIII and thus participate in the regulation of clotting.

There are several chemically different compounds which may interact with the thrombin molecule and inhibit the clotting ability of the latter protein. The so-called true antithrombin is a plasma protein. It has been purified to a great extent but its physicochemical characterization is still incomplete. True antithrombin migrates as an α globulin in electrophoresis. It can be adsorbed to aluminum hydroxide or barium sulfate gels. This step alone results in a five- to tenfold purification. Antithrombin destroys the thrombin molecule in vitro and in vivo. The physiological importance of this factor in regulating coagulation has been questioned.

Heparin. Heparin is a mucopolysaccharide $(C_{24}H_{31}O_{35}N_2S_5Na_7)$ and its molecular weight falls within the range of 12,000 to 45,000. The anticoagulant action of heparin is lost in weak acids, concomitant to the appearance of free amino groups on the heparin molecule. There is some evidence that heparin needs a plasma protein cofactor for its anticoagulant activity. Some investigation suggests that true antithrombin and the heparin cofactor are the same proteins. The main anticoagulant activity of heparin is the inhibition of thrombin-fibrinogen interaction. Besides this, some heparin inhibition of the prothrombin-thrombin conversion is also documented.

Some large-molecular-weight split products of fibrinogen inhibit clot formation. They may form as the effect of fibrinolysins* on the fibrinogen molecule. Besides inhibiting the fibrin polymerization process some of them inhibit the thrombin-fibrinogen interaction.

It was pointed out previously that Ca^{++} participates in several steps of the coagulation process. Substances capable of reducing the Ca^{++} concentration of blood will act as potent anticoagulants (EDTA, citrate, oxalate, and so forth).

Some essential factors of the coagulation process readily adsorb to such inorganic gels as $Al(OH)_3$ and $BaSO_4$. Therefore adsorbed blood plasma also becomes incoagulable.

Recently a new class of potential inhibitors of clotting has been discovered by Lorand and his collaborators. They have shown that the last step in the solidification of fibrin involves a transamination resulting in cross-link formation between the fibrin molecules. These cross-links are formed between an amino group of one fibrin monomer molecule and a carboxyl group of the other. For this reason Lorand et al. investigated a large number of synthetic amines as candidates for interaction with protein carboxyl side chains, inhibiting thereby the carbonyl amide bond formation

---

* Fibrinolysins are enzymes which hydrolyze fibrin.

that is the cross-linking process. Among the amino compounds, they synthesized those which, containing pentamine or hexamine residues attached to an apolar group toluene or naphthalene, were potent inhibitors. Best inhibition of cross-linking was obtained with N-(aminopentyl)-5-dimethyl-amino-1-naphthalene sulfonamide:

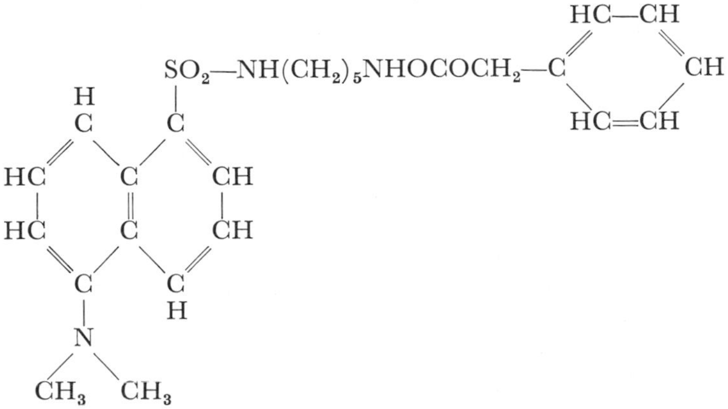

Fibrinolysis

In the previous section we enumerated several naturally occurring or artificially synthesized compounds which are capable of suppressing the coagulation process and thereby preventing the formation of blood clots. The plasma and several other tissues of the mammalian organism contain substances which are able to dissolve clots once they are formed. The potential biological and medical importance of these compounds is great. Fibrin clots may obstruct important blood vessels or cause adhesions between neighboring surfaces of different organs. The activity of the fibrinolytic substances in the body may reopen the closed blood vessel and terminate pathological connections between organs. On the other hand, increased fibrinolytic activity in the plasma may dissolve fibrin while it is being formed and therefore inhibit the formation of a solid durable clot which is necessary to stop bleeding from injured vessels. Needless to say, such overactivity of the fibrinolytic enzymes may prolong bleeding. Plasminogen-plasmin appears to be the most important representative of the fibrinolytic enzymes.

Plasminogen and Plasmin. Plasmins are proteolytic enzymes which are present in the circulating blood plasma in an inactive form called plasminogen. An alternative name for plasmin is fibrinolysin. Partially purified plasminogen is available with a 400 fold increase of its specific activity. In electrophoresis it behaves as α globulin; on starch block plasminogen migrates between β and α globulin. Physical measurements have shown it to have a molecular weight of 84,000 or 143,000 and an asymmetrical molecular shape. Plasminogen is believed to be a glycoprotein which also contains small amounts of phosphorus. The chemical properties of plasmin are similar to those of plasminogen, although the molecular weight of plasminogen is probably higher. There is some evidence that plasmin splits preferentially arginyl and lysyl bonds. The pH optimum for hydrolytic activity is between 7.0 and 8.0. Unlike plasminogen, plasmin is very unstable.

The plasmin activity of blood plasma, but not necessarily of the "purified" plasminogen preparations, may be activated by organic solvents. The mechanism of this activation is not clear. Another potent activator of plasminogen is streptokinase. This hydrolytic enzyme, obtained from cultures of *Streptococcus hemolyticus*, was thought to possess fibrinolytic activity itself. It was, however, demonstrated that streptokinase dissolves fibrin clots only in the presence of a plasma factor present in the euglobulin fraction. This plasma factor turned out to be plasminogen. Unfortunately the mechanism by which streptokinase activates plasminogen is not known.

Whether plasmin is a single enzyme or a group of enzymes is not known. Plasmin can digest a variety of proteins. From our point of view its action on clotting factors is the most interesting. Fibrin hydrolysis leads to formation of high-molecular-weight products, some of which possess anticoagulant activity. Besides fibrinogen, other clotting factors such as Factor VIII, Factor VII, and Factor V are among the proteins which may be hydrolyzed by plasmin.

Death or severe hemorrhages associated with increased plasmin activity are well documented in the clinical literature. Although the clinical conditions connected with such bleeding show widely divergent etiology, increase of fibrinolysis is well documented in connection with such clinical conditions as complicated childbirth, major surgery, and liver disease.

The Role of Vitamin K in Coagulation of Blood

The vitamin K family includes a number of naturally occurring or artificially synthesized α-naphthoquinone compounds. The basic structure of these vitamins is present in vitamin K_3:

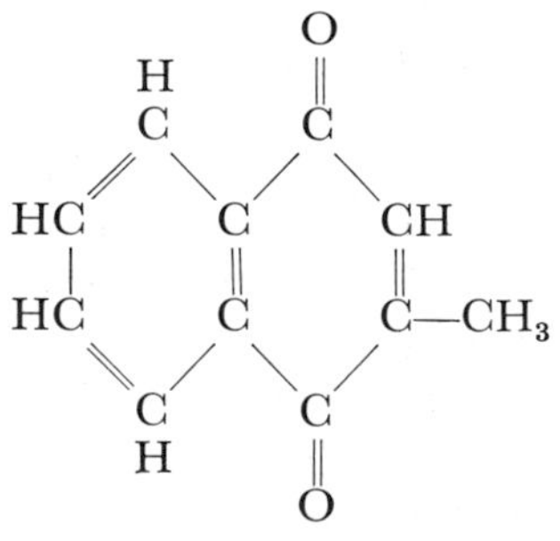

2-Methyl-1,4-naphthoquinone

Vitamin K is needed for the synthesis of prothrombin and Factors VII, IX, and X. Dicumarol is a structural analog of vitamin K. If Dicumarol is administered to man or many other mammalian organisms the prothrombin and factor VII concentrations of the plasma decline. Increased administra-

3,3′-Methylenebis(4-hydroxycoumarin)

tion of vitamin K will reverse the Dicumarol effect and reinstate normal prothrombin concentration in the plasma.

The molecular mechanism of action of vitamin K is not known. It is firmly established that this vitamin is not incorporated into the structure of prothrombin, nor does it serve as a cofactor for prothrombin. The antagonism between Dicumarol and vitamin K is used in medical therapy to suppress blood coagulation in certain diseases associated with thrombosis.

PHYSIOLOGICAL CHEMISTRY OF MUSCULAR CONTRACTION

The macromolecular chemistry of muscular contraction began with the observation of Engelhardt and Ljubamova that the the residue of skeletal muscle tissue, after being extracted by a dilute buffer solution, changes its shape upon the addition of ATP and at the same time hydrolyzes ATP to ADP and P_i. This experiment implicated the muscle residue as the source of contractile proteins and the ATP as the energy supply or trigger for the in vitro "contraction." From this muscle residue, Szent-Györgyi, Straub, and their colleagues isolated two proteins, myosin A and actin, which are the most important protein components of the contractile mechanism.

From physicochemical and electron microscopic measurements, myosin A appears to be a rodlike protein with a molecular weight of 500,000 (Fig. 8-14). In the neutral pH range it is soluble only in the presence of high salt concentration (0.6 M KCl). Myosin A hydrolyzes ATP to ADP and P_i. This activity is enhanced by Ca^{++} and inhibited by Mg^{++}.

Actin is soluble in dilute salt solutions and in vitro may have two different forms. In the presence of 0.1 M KCl and some Mg^{++} the actin solution is viscous and the actin molecules are polymerized. This form of actin is known as F-actin (fibrillar) and there is good evidence that F-actin is present in the muscle tissue in vivo. In the absence of salts the F-actin depolymerizes to its components which are known as G-actin (globular) molecules and have

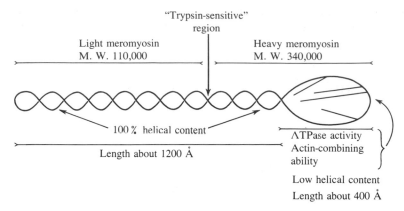

Figure 8-14.   Some molecular parameters of myosin A.

a molecular weight of 60,000.   One molecule of G-actin binds one molecule of ATP, and during the G-F transformation (polymerization) this G-actin-bound ATP is dephosphorylated so that the F-actin contains one molecule of ADP per molecule of G-actin subunit.   The F-actin extracted from the muscle tissue also contains a similar amount of firmly bound ADP.   Except for this weak ATP splitting observed in the polymerizing G-actin solution, the actin alone has no ATPase activity.

A remarkable quality of actin is its ability to combine with myosin. This actin-myosin complex, often called the actomyosin molecule, is the smallest unit which under suitable experimental conditions may serve as a muscle model in studies of contraction.

Further studies on the molecular structure of myosin A have shown that trypsin splits this molecule into two parts (see Fig. 8-14).   These two components of the myosin A molecule have different sedimentation velocities in the ultracentrifuge.   The one that sediments faster is called heavy meromyosin and the other, light meromyosin.   It is established that heavy meromyosin retains the ATPase activity and the actin-combining ability of the myosin A molecule, but, unlike myosin A, this fragment becomes soluble in dilute salt solutions.   Light meromyosin retains the solubility characteristics of the parent myosin A molecule but has neither ATPase nor actin-combining ability.   Electron microscropic and hydrodynamic observations indicate that heavy meromyosin has a globular part with a tail attached to it whereas light meromyosin is a long thin rodlike molecule.

In the actomyosin complex the myosin A is attached to the actin by its heavy meromyosin segment.   The kinetic parameters of the actomyosin ATPase are different from those of the myosin A ATPase.   One important difference is that Mg^{++}, which is inhibitory for the myosin A ATPase, is an activator for the actomyosin ATPase.   Actin therefore may be considered a modifier of the myosin A enzyme.

In the presence of low salt concentration, small amounts of Mg^{++}, and trace amounts of Ca^{++}, the actomyosin suspension becomes more turbid upon the addition of a suitable amount of ATP. The actomyosin particles become more dense, and they shrink and precipitate. This effect of ATP on the actomyosin suspension was called "superprecipitation" by Szent-Györgyi and it is widely used to study the molecular mechanism of muscular contraction in vitro.

Electron microscopic investigations of Huxley and others produced important results which have had far-reaching influence on the thinking of biochemists and physiologists who are interested in the molecular mechanism of muscular contraction.

It may be seen from Figure 8-15 that from the morphological point of view the muscle fiber may be subdivided into small units called sarcomeres (from Z line to Z line). Myosin A (thick lines) is located in the A band; actin (thin line) starts at the Z line and ends at the H zone within the A band. The I band contains only F-actin filaments. The A band of the resting muscle may be subdivided longitudinally into three approximately equal parts. The section in the middle is called the H zone and contains only myosin A filaments. The two lateral sections of the A band contain both myosin A and actin filaments. This is significant because it proves morphologically that actin and myosin interdigitate in certain sections of the sarcomere. Physical contact between the actin and myosin is ensured by cross-bridges which are a part of the myosin molecules and extend toward the actin filament.

From electron microscopic and biochemical evidence it is widely accepted that a segment of the globular heavy meromyosin part of the myosin A molecule is involved in the cross-bridge formation. According to Huxley, the fine structure of the myosin filament in the A band may be explained as

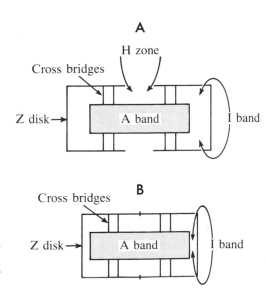

Figure 8-15. Diagrammatic illustration of a sarcomere. *A*, Relaxed muscle. *B*, Contracted muscle.

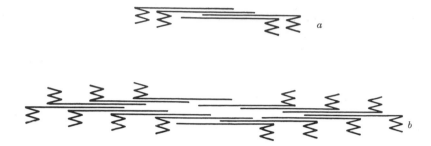

Figure 8-16. Preferential aggregation of myosin A molecules. *a*, Possible arrangement of myosin molecules, with globular region at one end only, to produce short filaments of the type observed, with globular projections at either end and a straight bare shaft in the center. The polarity of the myosin molecules is reversed on either side of the center. *b*, Possible arrangement of same myosin molecules to produce longer filaments in which the bare shaft in the center is still present, but in which a longer region on either side now has globular projections on it. The polarity of the myosin molecules is reversed on either side of the center, but all the molecules on the same side have the same polarity. (From Huxley, H. E.: Structural arrangements and the contraction mechanism in striated muscle. Proc. Roy. Soc. Biol. *160*:442–448, 1964.)

an aggregate of myosin A molecules. As depicted in Figure 8-16, the myosin A molecules show a side to side aggregation in staggered arrangement.

During contraction the I band becomes shorter, the width of the A band remains unchanged, but the H zone within the A band becomes narrower or disappears (Fig. 8-15*B*). Using these findings, Huxley proposed the sliding filament theory for muscular contraction which enjoys widespread support among physiologists and biochemists at this time. During contraction the actin filaments slide toward the center of the sarcomere into the H zone. Since the H zone becomes narrower or disappears, the sarcomere may shorten without the concomitant shortening of either the myosin or the actin filament. As Huxley visualizes the process, during contraction ATP is being hydrolyzed on the cross-bridges and the liberated energy somehow transforms the cross-bridge to an impelling device which moves the actin filament. It must be emphasized at this point that this theory does not deal with the nature of the mechanochemical coupling.

In my opinion the nature of the molecular forces responsible for the relative actin to myosin displacement during contraction have yet to be delineated. Moreover the chemical mechanism of the cyclic association and dissociation of the moving actin molecules with the "nonmoving" myosin molecules remains to be experimentally established.

Morales and Botts proposed a different biochemical mechanism to explain muscular contraction. Their theory assumes that the myosin A molecules located on the sleeve of the A band are linked to actin in both the contracting and the relaxed muscle, so that the mechanical continuity between myosin A and actin is always preserved. A further assumption is that the myosin A molecules in the sleeve of the A band are arranged around a central core which may contain some protein, probably other than myosin A (e.g., tropomyosin). During contraction the state of aggregation of the myosin A molecules changes in such a way that the longitudinal distance between the myosin A molecule decreases (this would be analogous to the tightened aggregation observed during superprecipitation), so that the actin filaments are being pulled inward between the myosin A filaments. Since these changes are restricted to the sleeve of the A band, the length of the central core may remain unchanged. Such a mechanism would not contradict the fact, observed in experiments, that the length of the A band does not decrease during contraction There is no need to assume the presence of a flexible joint within the myosin A molecule since the state of aggregation of the myosin A molecules is changing during contraction.

The theory of Morales and Botts also proposed a mechanism for the ATP effect on the molecular level. The resting length of the muscle in vivo, as well as the resting length of the actomyosin suspensions in vitro, is supposed to be determined by an equilibrium between the forces which promote aggregation of the myosin A molecules and the repelling forces due to the electrostatic repulsion of the similarly charged groups on the myosin A molecules. The forces that promote aggregation of myosin A molecules are unknown, a likely possibility being electrostatic attraction between oppositely charged groups. In either case the adsorption of ATP to specific myosin A sites could discharge the molecule and shift the existing equilibria in favor of the attracting forces, producing superprecipitation in vitro and contraction in vivo. In the meantime ATP is being split by the contracting muscle. The affinity of ADP to the myosin A binding sites has been shown to be much smaller than that of ATP. Therefore the ADP dissociates from the myosin A molecules and the equilibrium between attractive and repelling forces is shifted back to the level that existed before contraction took place. Figure 8-17 depicts a muscle model constructed by Morales to visualize the mechanism of contraction which has just been discussed. The operation of the model is explained in the legend.

One difficulty with this theory is that the available x-ray diffraction data do not indicate the presence of any major molecular rearrangement within either the myosin A or the actin filament during contraction. Since according the theory of Morales the state of aggregation of myosin A molecule is changing during contraction, molecular rearrangements within the A band would be expected to occur.

Both Huxley's and Morales' theories respect the fact that during physiological muscular contraction the length of the myosin and actin filaments remains unchanged. Huxley postulates that contraction proceeds without any molecular rearrangement among the myosin A molecules within the myosin A filament, while Morales assumes that such changes have to occur. Both theories agree that at least a part of the myosin A molecule undergoes some conformational change during contraction but neither one deals with the

Figure 8-17. Model of a portion of a thick filament and six neighboring thin filaments (straight steel rods); left, relaxed; right, contracted. In this model, a single myosin molecule is a bent white rod arranged so that the bulbous (heavy meromyosin) end projects radially to touch an actin filament, and the shaft (light meromyosin) portion lies almost parallel to the thick filament axis. The bent points of the myosin molecules lie on a helix whose diameter and pitch conform with observed values. At the center of the thick filament is a cylindrical "core" (Plexiglass tube). Contraction is supposed to occur because the array of myosin molecules around the core is rearranged in such a way that the bend points now lie on a helix of reduced pitch and correspondingly increased diameter; the right-hand model illustrates a contraction of about one-third. After the rearrangement, the myosin shafts have been further telescoped into one another (aggregated in an orderly manner), and the translation of the bulb projections has towed the actin filaments. The core has remained of the same length. (From Morales, M. F.: On the Mechanochemistry of Contraction. *In* Pullman, B., and Weissbluth, M. (eds.): Molecular Biophysics. New York, Academic Press, 1965, pp. 397–410.)

possibility that actin also undergoes some cyclic conformational change during the contraction-relaxation process. In this respect A. G. Szent-Györgyi recently brought forward some evidence that during in vitro contraction (superprecipitation) actin is involved in a cyclic conformational change which may play a role in the actin-myosin cross-linking during contraction.

It is firmly established that actin, myosin, Mg^{++}, Ca^{++}, and ATP are essential components of the contractile machinery. This does not exclude the possibility that other known or unknown compounds may also participate in the process. The physical interaction between actin and myosin is essential for contraction, and contraction proceeds without any major change in the length of the actin and myosin filaments, but to what extent protein conformational change is necessary for contraction is not clear. The precise mechanism of the mechanochemical coupling is unknown.

PEPTIDE AND PROTEIN HORMONES

Metabolic processes in higher organisms are regulated by compounds (hormones) produced in the cells of endocrine glands. Hormones are secreted into the blood which carries them to the tissue or tissues (target organ) where they exert a specific effect. A large number of the important hormones are proteins or polypeptides. They represent a class of proteins the role of which is to regulate fundamental chemical and physical processes such as growth, reproduction, metabolism of nutrients, and transport across cellular membranes in the mammalian organism. One of these endocrine glands, the hypophysis (or pituitary), is often called the "master gland." This name was given because practically all other endocrine glands in the body are under the influence of the hypophyseal hormones. On the other hand, the hypophysis itself is under the influence of the nervous system. Figure 8-18 shows a simplified scheme of this correlation.

Releasing Factors

The peptide hormones produced in the central nervous system are commonly known as the releasing factors. Upon incubation with a specific

Figure 8-18. Functional correlation between the nervous system, the hypophysis, and the other endocrine glands. The releasing factors secreted in the nervous system influence hormone release from the hypophysis. Hypophyseal hormones act on many target glands. The hormones produced by the target glands may in turn influence hormone production in the hypophysis.

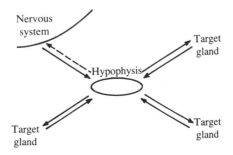

CyS–Tyr–Ileu–Glu(NH$_2$)–Asp(NH$_2$)–CyS–Pro–Leu–Gly(NH$_2$)
1   2   3   4   5   6   7   8   9

Figure 8-19. Oxytocin.

releasing factor the hypophysis releases a specific hormone; e.g., hypophysis incubated with corticotropin-releasing factor will release corticotropin. The mechanism of action of these releasing factors is not known. They may stimulate hormone synthesis in the hypophysis or facilitate the release of the intracellular hormone into the blood. The structure of the corticotropin-releasing factor is tentatively established. There is some experimental basis to assume the existence of the following other releasing factors: luteinizing hormone-releasing factor, follicle-stimulating hormone-releasing factor, and thyrotropin-releasing factor.

Hypophyseal Hormones

OXYTOCIN AND VASOPRESSIN.   Oxytocin and vasopressin cause contraction of smooth muscles, oxytocin's effect being strong on uterine muscle and vasopressin's on arterial muscle. Vasopressin also has antidiuretic action. Oxytocin and vasopressin are isolated from the neurohypophysis (posterior lobe of the pituitary gland). Their structure is quite similar (Figs. 8-19 and 8-20). The main difference is that at position 8 oxytocin has a leucine and vasopressin has either an arginine or a lysine molecule.

The size and intactness of the ring structure between amino acids 1 and 6 is important for the biological activity; reduction of the —S—S— linkage between 1 and 6 abolishes biological activity. The basicity of the amino acid in position 8 favors oxytocic activity.

CORTICOTROPIN.   Corticotropin, which is produced in the basophil cells of the adenohypophysis (anterior part of the hypophysis), promotes the formation of corticosteroids in the adrenal gland (Fig. 8-21). It has also direct effect on the metabolism of carbohydrates, fats, and proteins. The purified hormone has a molecular weight of 4500. The biological activity of the hormone is destroyed by oxidation (H$_2$O$_2$) or periodate.

CyS–Tyr–Phe–Glu(NH$_2$)–Asp(NH$_2$)-CyS–Pro–Arg–Gly(NH$_2$)
1   2   3   4   5   6   7   8   9

Arginine vasopressin

CyS–Tyr–Phe–Glu(NH$_2$)–Asp(NH$_2$)–CyS–Pro–Lys–Gly(NH$_2$)
1   2   3   4   5   6   7   8   9

Lysine vasopressin

Figure 8-20. Vasopressin.

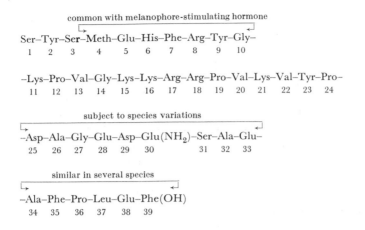

common with melanophore-stimulating hormone

Ser–Tyr–Ser–Meth–Glu–His–Phe–Arg–Tyr–Gly–
1   2   3   4   5   6   7   8   9   10

–Lys–Pro–Val–Gly–Lys–Lys–Arg–Arg–Pro–Val–Lys–Val–Tyr–Pro–
11 12 13 14 15 16 17 18 19 20 21 22 23 24

subject to species variations

–Asp–Ala–Gly–Glu–Asp–Glu(NH$_2$)–Ser–Ala–Glu–
25 26 27 28 29 30   31 32 33

similar in several species

–Ala–Phe–Pro–Leu–Glu–Phe(OH)
34 35 36 37 38 39

Figure 8-21. Amino acid sequence of human corticotropin.

A suggested mechanism of action of corticotropin on the adrenal gland is the initiation of a series of reactions which ultimately promote the formation of steroids.

The amino acid sequence between positions 4 and 10 is common with that of the melanophore-stimulating hormone and accounts for the melanocyte-stimulating activity of this hormone. This sequence is invariant in several mammalian species. The amino acid sequence between positions 25 and 33 shows variation from species to species. The amino acid sequence between 1 and 24 is important for the biological activity of corticotropin.

MELANOPHORE-STIMULATING HORMONE (MSH). Excessive production of MSH is probably responsible for the darkening of skin in some pathological conditions. The source of MSH is the adenohypophysis. It consists of 13 amino acids.

GROWTH HORMONE (GH). GH has a well documented effect on the growth and development of the organism. The mechanism of action of this hormone on cellular and subcellular level is not well understood, but there is some evidence that GH has a stimulating effect on RNA and protein synthesis. It is extracted from the pituitary gland. The molecular weight of GH obtained from various sources is different; human GH has a molecular weight of 24,000 while the molecular weight of bovine GH is close to 45,000.

PROLACTIN. The most important biological effect of prolactin is the initiation and maintenance of lactation. The molecular weight of this hormone appears to be between 23,000 and 25,000.

FOLLICLE-STIMULATING HORMONE (FSH) AND LUTEINIZING HORMONE (LH). FSH and LH are produced by the basophilic cells of the anterior pituitary. Their biological role is to promote the development of the ovum in the ovaries, and they also have important functions during pregnancy. The precise composition of FSH and LH is not known because these hormones have not been isolated in completely pure form. FSH and LH both contain carbohydrates which appear to be essential to their activity. An interesting finding was that these hormones contain about 9 moles of cysteine residues per mole of protein, 8 of which appear to be involved in —S—S— bridges. Reduction of the —S—S— linkages leads to complete loss of activity. Apparently the four —S—S— linkages per mole of protein stabilize these proteins in a rigid nonhelical structure which is essential for their activities. Digestion of FSH and LH with trypsin, chymotrypsin, and pronase diminishes or abolishes (at 60 per cent digest) the biological activity of these hormones. Similarly, takadiastase also destroys the biological effect of FSH and LH; this suggests that the carbohydrate moiety is essential for the hormone activity.

THYROID-STIMULATING HORMONE (TSH). Strong evidence suggests that the basophil cells of the anterior pituitary are the source of TSH. The main biological function of TSH is to promote the generation of thyroid hormone. TSH is aglyco protein. Since at the present time there is no stable highly purified TSH available, the chemical structure is not known accurately. The molecular weight is estimated to be 26,000 to 30,000. The cysteine content is high but there is no evidence of —S—S— linkages. The carbohydrate moiety is essential for the biological activity.

Parathyroid Hormone

The biological function of this hormone is regulation of calcium and phosphate metabolism. It promotes the mobilization of calcium from the bones, increasing the blood calcium level, and it also enhances the excretion of phosphate from the body. The cellular or subcellular mechanism of action of parathyroid hormone is not completely understood. There is, however, good evidence that it acts on the bones, kidneys, the gastrointestinal tract, and lactating mammary glands. The source of the hormone is the parathyroid cells. The molecular weight of this peptide varies between 3800 and 8500. Parathyroid hormone and similar substances all contain 2-methionine, which is essential for their biological activity.

Calcitonin. The mean biological activity of calcitonin is opposite that of parathyroid hormone. It decreases the serum calcium level, and it acts much faster than parathyroid hormone. Calcitonin probably originates in the so-called "mitochondria-rich" or parafollicular cells of the thyroid gland (thyrocalcitonin). Calcitonin is either a polypeptide or a low-molecular-weight protein the biological activity of which is completely destroyed by pepsin.

Pancreatic Hormones

INSULIN. The biological effect of insulin is to enhance the uptake of glucose by the tissues. The precise mechanism of action is still debated. Insulin is produced in the β cells of the pancreas. The elucidation of its chemical structure is one of the major achievements of biochemistry in this century (Sanger) (Fig. 8-22).

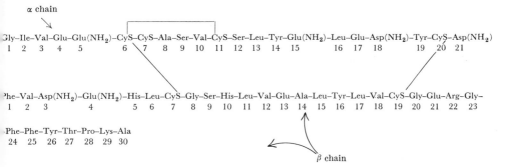

Figure 8-22. Amino acid sequence of insulin.

The intra- and interchain disulfide bridges, as well as the amino acid sequence 19 to 21 in chain α (Tyr-Cys-Asp), are essential for the biological activity of insulin. Masking more than two free amino groups, or iodination of tyrosine 26 in the β chain destroys activity. On the other hand, the removal of the first six N terminal residues from the β chain by leucine aminopeptidases does not inactivate the hormone.

Some experimental results suggest the presence of insulin-inhibitory factor in the blood. A polypeptide called "synalbumin" found in increased amount in the blood of diabetics is described as an insulin antagonist. There is some evidence that this inhibitory factor may be identical to the β chain of insulin which under certain conditions may be attached to albumin and in that form inhibits insulin effect. The sulfonated and reduced β chain of insulin has a small but consistent inhibitory effect on the parent hormone. The mechanism of action of the insulin inhibition is not known.

GLUCAGON. This compound was discovered as an impurity attached to early insulin preparations. Its effect is in many respects antagonistic to insulin. Administration of glucagon will increase rather than lower the serum glucose concentration. Glucagon therefore is also called "hyperglycemic factor." It is produced in the α cells of the pancreas. Its molecular weight is 3550, and it contains 29 amino acid residues but no sulfhydryl groups (Fig. 8-23).

His–Ser–Glu(NH$_2$)–Gly–Thr–Phe–Thr–Ser–Asp–Tyr–
  1   2   3   4   5   6   7   8   9   10

–Ser–Lys–Tyr–Leu–Asp–Ser–Arg–Arg–Ala–Glu(NH$_2$)–
 11   12   13   14   15   16   17   18   19   20

–Asp–Phe–Val–Glu(NH$_2$)–Trypt–Leu–Meth–Asp(NH$_2$)–Thr.
 21   22   23   24   25   26   27   28   29

Figure 8-23. Amino acid sequence of glucagon.

REFERENCES

Transport and Regulatory Functions of Proteins

1. Laurell, C. B.: Metal-binding plasma proteins and cation transport. *In* Putnam, F. W. (ed.): The Plasma Proteins. Vol. 1. New York, Academic Press, 1960, pp. 349–378.
2. Putnam, F. W.: Structure and function of the plasma proteins. *In* Neurath, H. (ed.): The Proteins. Vol. 3. New York, Academic Press, 1965, pp. 153–267.

Immunochemistry

1. Boyd, W. C.: Fundamentals of Immunology. 4th edition. New York, Interscience Publishers, 1967.
2. Edelman, G. M.: Molecular mechanisms of the immune response. *In* Hayashi, T., and Szent-Gyorgyi, A. (eds.): Molecular Architecture in Cell Physiology. Englewood Cliffs, N.J., Prentice-Hall, Inc., 1966, pp. 99–121.
3. Haber, E.: Immunochemistry. Ann. Rev. Biochem. *37:* 497–520 (1968).
4. Haurowitz, F.: Immunochemistry and the Biosynthesis of Antibodies. New York, Interscience Publishers, 1968.
5. Singer, S. J.: Structure and function of antigen and antibody proteins. *In* Neurath, H. (ed.): The Proteins. Vol. 3. New York, Academic Press, 1965, pp. 269–357.

The Coagulation of Blood

1. Davie, E. W., and Ratnoff, O. D.: The proteins of blood coagulation. *In* Neurath, H. (ed.): The Proteins. Vol. 3. New York, Academic Press, 1965, pp. 359–443.
2. MacFarlane, R. G.: The Blood Coagulation System. *In* Putnam, F. W. (ed.): The Plasma Proteins. Vol. 2. New York, Academic Press, 1960, pp. 137–181.
3. Miale, J. B.: Laboratory Medicine—Hematology. St Louis, C. V. Mosby Co., 1967.
4. Seegers, W. H. (ed.): Blood Clotting Enzymology. New York, Academic Press, 1967.
5. Tocantins, L. M., and Kazal, L. A. (eds.): Blood Coagulation, Hemorrhage and Thrombosis. New York, Grune and Stratton, Inc., 1964.

Muscle Contraction

1. Gergely, J. (ed.): Biochemistry of Muscle Contraction. Boston, Little, Brown & Co. 1964.
2. Gergely, J.: Contractile proteins. Ann. Rev. Biochem. *35:* 691–722 (1966).
3. Huxley, H. E.: Structural evidence concerning the mechanism of contraction in striated muscle. *In* Paul, W. W., Daniel, E. E., Kay, C. M., and Monckton, G. (eds.): Muscle. New York, Pergamon Press, 1965, pp. 3–28.
4. Morales, M. F.: Some general features of the mechanism of muscle contraction. *In* Mazia, D., and Tyler, A. (eds.): General Physiology of Cell Specialization. New York, McGraw-Hill Book Co., 1963, pp. 266–276.
5. Morales, M. F.: On the mechanochemistry of contraction. *In* Pullman, B., and Weissbluth. M. (eds.): Molecular Biophysics. New York, Academic Press, 1965, pp. 397–410.
6. Szent-Gyorgyi, A. G.: The role of actin-myosin interaction in contraction. *In* Aspects of Cell Motility. XXII Symposium of the Society for Experimental Biology. Cambridge, Cambridge University Press, 1968, pp. 17–42.

Protein Hormones

1 Butt, W. R.: Hormone Chemistry. Princeton, N.J., D. Van Nostrand Co., 1967.
2. Recent Progress in Hormone Research. New York, Academic Press.

9

METABOLISM
OF PROTEINS

DIGESTION OF PROTEINS IN THE GASTROINTESTINAL
TRACT

The physical grinding and mixing movements of the stomach work like a homogenizer; they break up the food particles and enlarge the contact surface between gastric juice and ingested food. The dietary proteins are first attacked by proteolytic enzymes in the stomach. The high acidity of the gastric juice denatures many proteins and so renders them more susceptible to enzymatic attack. As a result of the combined action of several proteolytic enzymes such as pepsin, parapepsin I, parapepsin II, and gastricsin, the dietary proteins which reach the duodenum (uppermost part of the intestinal tract) become a mixture of mainly proteins (mostly denatured and partly digested), polypeptides, and a much smaller quantity of smaller peptides and amino acids.

This mixture is rapidly hydrolyzed in the upper intestinal tract by the various pancreatic and intestinal proteases—trypsin, chymotrypsin, elastase, carboxypeptidases, aminopeptidases, and dipeptidase. It has been determined that after a 500 gm. standard meal containing 32 gm. protein, the intestinal juice in which this meal is suspended or dissolved is about 2000 ml. The concentration of the individual proteolytic enzymes varies between 200 and 800 μg. per ml. so that 0.1 to 1.0 gm. enzyme is available for the digestion of the 32 gm. dietary protein. The high enzyme concentration in the upper intestinal tract and the fast removal of the hydrolytic products account for the fast and efficient proteolysis in the gut.

There is some uncertainty regarding the absorption and breakdown of small peptides, however. There is no unequivocal evidence that the breakdown of proteins to amino acids is completed in the intestinal lumen. It

Table 9-1. *Proteolytic Enzymes*

INACTIVE PROENZYME (ZYMOGEN)	TRANSFORMING PRINCIPAL	ACTIVE PROTEOLYTIC ENZYME	MODE OF ACTION	pH OPTIMUM	PEPTIDE BONDS ATTACKED	LOCATION OF SECRETION
Pepsinogen	Strongly acidic pH	Pepsin	Endopeptidase	Acidic	Adjacent to aromatic amino acids	Chief cells of stomach
Trypsinogen	Enterokinase and a small amount of trypsin liberated by the enterokinase (autocatalytic)	Trypsin	Endopeptidase	Slightly alkaline	Adjacent to lysine or arginine	Pancreas
Chymotrypsinogen		Chymotrypsin	Endopeptidase	Slightly alkaline	Adjacent to aromatic amino acids	Pancreas
Proelastase	Trypsin	Elastase	Endopeptidase	Slightly alkaline	Adjacent to neutral aliphatic amino acids	Pancreas
Procarboxypeptidase A	Trypsin	Carboxypeptidase A	Exopeptidase	Slightly alkaline	Liberates amino acids in carboxyl terminal position. Preferentially hydrolyzes branched chains and aromatic amino acids in carboxyl terminal position	Pancreas
Procarboxypeptidase B	Trypsin	Carboxypeptidase B	Exopeptidase	Slightly alkaline	Liberates basic amino acids in carboxyl terminal position	Pancreas
—	—	Leucine aminopeptidase	Exopeptidase	Slightly alkaline	Liberates amino terminal amino acids, preferentially leucine or aliphatic amino acids in amino terminal positions	Intestinal mucosa
—	—	Dipeptidase	—	—	Dipeptide	Intestinal mucosa

may very well be true that the site of action of some exopeptidases and di-peptidases is the brush border of the epithelium or perhaps the inside of the epithelial cell. On the other hand, it is experimentally well established that under normal conditions (except for a few days after birth) neither protein nor peptides are transported through the intestinal cells to the circulation.

Mechanism of Protein Hydrolysis in the Gastrointestinal Tract

Many proteolytic enzymes are secreted into the gastrointestinal tract in their enzymatically inactive form (zymogen). These inactive proenzymes are transformed to their active forms by a hydrolytic process which involves the loss of a peptide or peptides from the proenzyme (see also Table 9-1). The active proteolytic enzymes are conveniently classified as endopeptidases and exopeptidases.

Endopeptidases attack peptide bonds within the polypeptide chain. This does not mean that endopeptidases may not hydrolyze terminal amino acids under certain conditions. It is, however, well documented that endopepti-dases preferentially hydrolyze more centrally located peptide bonds of the peptides and proteins (see also Table 9-1).

Exopeptidases attack the peptide bond adjacent to the free carboxyl end (carboxyl terminal) or amino end (amino terminal) of peptides and pro-teins. Carboxypeptidases liberate amino acids from the carboxyl terminal position of peptides and proteins, while aminopeptidases split off the amino terminal amino acids from peptides and proteins (see also Table 9-1).

The hydrolysis of proteins in the gastrointestinal tract is a combined action of various endopeptidases and exopeptidases. This process is sche-matically illustrated in Figure 9-1. First the proteins are hydrolyzed mainly to polypeptides and only small amounts of amino acids are liberated. This is due to the fact that on the protein molecule only a relatively small number of carboxyl terminal and amino terminal groups are available for the exopeptidases but a much greater number of peptide bonds are sensitive to the attack of endopeptidases (A in Figure 9-1). Once a protein molecule is broken down to several polypeptides the exopeptidases have more free carboxyl and amino end groups to work on and so their contribution to the hydrolytic process increases (B in Figure 9-1). While the hydrolysis of the protein molecule proceeds an increasing number of peptide bonds become available for the endopeptidases. These newly exposed peptide groups are "protected" from the endopeptidases until the native protein structure is preserved. The polypeptides therefore will be cleaved further to yield a large number of smaller peptides and an increased amount of amino acids.

A   B

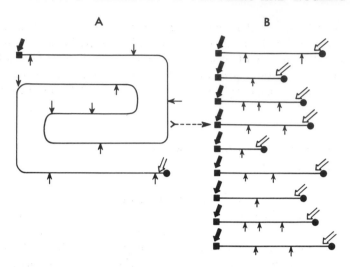

Figure 9-1.   Schematic illustration of proteolysis as the combined action of endopeptidases and exopeptidases. *A* = Protein X. *B* = Protein X hydrolyzed to nine polypeptides. ↑↓ = Bonds hydrolyzed by various endopeptidases. ↓ = Bonds hydrolyzed by aminopeptidases (exopeptidases). ⇓ = Bonds hydrolyzed by carboxypeptidases (exopeptidases). ■ = Amino terminal. ● = Carboxyl terminal.

This process repeats itself on the smaller peptides until amino acids become the predominant end product of the hydrolytic process. The hydrolysis of proteins in the gastrointestinal tract may be summarized as follows:

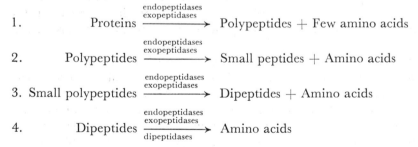

1.   Proteins $\xrightarrow{\substack{\text{endopeptidases} \\ \text{exopeptidases}}}$ Polypeptides + Few amino acids

2.   Polypeptides $\xrightarrow{\substack{\text{endopeptidases} \\ \text{exopeptidases}}}$ Small peptides + Amino acids

3. Small polypeptides $\xrightarrow{\substack{\text{endopeptidases} \\ \text{exopeptidases}}}$ Dipeptides + Amino acids

4.   Dipeptides $\xrightarrow{\substack{\text{endopeptidases} \\ \text{exopeptidases} \\ \text{dipeptidases}}}$ Amino acids

ABSORPTION OF AMINO ACIDS FROM THE INTESTINES

The amino acids produced during the digestion are absorbed into the intestinal cell and transported to the portal vessels. Until recently it was believed that amino acid absorption in the gut could be explained on the basis of simple diffusion. Many experimental results obtained in the last 18 years, however, have led to different conclusions. It has been shown that many amino acids are absorbed against a concentration gradient and that this process may be inhibited by metabolic poisons. Amino acids belonging to the L series are absorbed faster from the intestines than their respective D enantiotropic forms. Sodium ions increase and potassium ions decrease the absorption of several amino acids. Many amino acids show competition for absorption. In view of all these facts it is widely accepted that the absorption of amino acids from the lumen of intestine involves active transport and cannot be explained on the basis of free diffusion.

Under normal conditions absorption of dietary amino acids appears to reach 90 to 95 per cent completion within the upper third of the jejunum; this leaves a large margin of safety for the mammalian organism since the lower two-thirds of the jejunum and the ileum are capable of absorbing amino acids efficiently.

AMINO ACID TRANSPORT

From intestinal-wall cells the amino acids enter the portal veins. After a meal containing enough protein to maintain the organism in nitrogen equilibrium,* the amino acid level of the portal vein rises conspicuously. The amino acid concentration of the arterial blood also is increased but not nearly so much as that of the portal vein because about 80 per cent of the incoming amino acids is taken up by the liver.

By way of the arterial circulation the amino acids reach every tissue and organ of the body and are taken up and used according to the metabolic needs of the various organs. Since under normal conditions the tubules of the kidneys reabsorb about 90 per cent of the amino acid content of the glomerular filtrate, only a very small amount of the amino acids appears in the urine.

As in the case of amino acid absorption, an important question in connection with amino acid transport is the nature of the transport process itself. Since amino acids are taken up by various tissues (muscle, liver, red blood cell, and so forth) against a concentration gradient, free diffusion alone cannot account for it. Other observations that indicate that active transport must be involved are similar to those mentioned in connection with absorption of amino acids. Amino acids belonging to the L series are usually taken up faster than their corresponding D enantiotropic forms. Various metabolic inhibitors such as dinitrophenol and iodoacetic acid either considerably slow down or completely inhibit amino acid accumulation in various tissues.

It has also been discovered that competition exists among amino acids during the transport process. The competition is strong among various neutral amino acids (e.g., glycine competes with alanine); on the other hand, amino acids containing ionizing side chains do not compete appreciably with neutral amino acids. There is, however, a competition between the non-neutral amino acids. This competition is most conspicuous between lysine and ornithine (both basic) and glutamic and aspartic acid (both acidic) and somewhat less effective between a basic and an acidic amino acid.

It has been proposed that there are two mechanisms by which amino

---

* An organism is in nitrogen equilibrium if the dietary intake of nitrogen is equal to the excretion of nitrogen from the body.

acids enter the cells. One is a "pump" type of mechanism. This is uni-directional and is capable of building up a steep concentration gradient. It is very sensitive to metabolic poisons and probably is influenced by insulin. The other transport mechanism works on basis of "exchange." It is supposed that the carrier transports amino acids in both directions, and the amino acid that is preferentially moved inward by the carrier need not be the same as the one that is carried in the opposite direction. This latter mechanism is prob-ably less sensitive to metabolic poisons and is not influenced by insulin. Alanine and glycine, for example, are taken up into the cells by the pump mechanism, lysine has greater affinity for the exchange mechanism, and methionine appears to be transported by both.

In view of these facts one may understand why the concentration of amino acids in the blood plasma is not a reliable measure of the amino acid concentration of the tissues. First of all, there need not be a strict correla-tion between the plasma concentration of an individual amino acid and its entry into the tissue. Furthermore, competition between amino acids will result in a relative increase in amino acid concentration in the plasma. The plasma concentration of the amino acid whose entry into tissue cells was inhibited is high, but the tissue cell concentration is low.

Hormones with a strong anabolic effect have been found to enhance amino acid transport into the cells. The injection of hydrocortisone into experimental animals increases the amino acid concentration in the liver cells. Similarly, growth hormone and insulin have been shown to enhance the accumulation of amino acids in muscle tissue. The connection between the metabolic processes in the cells and the entry of amino acids into the tissues is not clear.

GENERAL OUTLINE OF THE INTERMEDIARY METABOLISM OF AMINO ACIDS

Free amino acids in the mammalian organism may become involved in the synthesis of new proteins or peptides, and they may undergo structural alterations and be transformed into hormones (epinephrine, thyroxine) or other biologically important compounds (serotonin, histamine). Many amino acids may be converted into carbohydrates or lipids, or may give rise to the formation of another amino acid. Amino acids are also catabolized to carbon dioxide, water, and urea, yielding energy during the process.

Since the amino acids have such diverse functions their metabolic path-ways are quite complicated. This chapter is intended merely to summarize some of the main routes for the degradation and synthesis of the most common amino acids. The interested reader should consult any detailed textbook on biochemistry for a more technical coverage of this important topic.

With respect to the catabolism of the various amino acids, the cleavage of the NH_2 group appears to be a common step. This may take place at various stages of the degradation process. A rather substantial body of evidence seems to indicate that transamination reaction plays a key role in this operation. Transamination means the transfer of an amino group from one molecule to another without the formation and involvement of free ammonia. It was found that α-ketoglutaric acid may serve as an amino group acceptor for most amino acids, and the enzyme which is able to catalyze the amino group transfer from amino acids to α-ketoglutarate is widespread in the tissues of the mammalian organism. This is one reason for the view that this enzyme, glutamic transaminase, is most important in the removal of the amino group from the amino acids. (In the subsequent discussion transaminase means glutamic transaminase.) The general reaction scheme of the transamination has been depicted in Chapter 6 and the importance of vitamin B_6 in this process has also been mentioned. Since most amino acids give rise to glutamic acid during their decomposition, there must be an effective mechanism to deaminate glutamic acid and make ketoglutaric acid available for further transaminations. Another enzyme, called L-glutamic acid dehydrogenase, which is widely distributed in almost all mammalian tissues, is able to catalyze such a reaction and thus participate in the decomposition of amino acids. The deamination of glutamic acid may be depicted as follows:

$$
\begin{array}{l}
\text{COOH} \\
|\\
\text{CH}_2 \\
|\\
\text{CH}_2 \qquad + \; \text{DPN}^+ \text{ (or TPN}^+) \; \underset{\text{dehydrogenase}}{\overset{\text{glutamic acid}}{\rightleftarrows}} \\
|\\
\text{CH—NH}_2 \\
|\\
\text{COOH} \\
\text{Glutamic acid}
\end{array}
$$

$$
\begin{array}{l}
\text{COOH} \\
|\\
\text{CH}_2 \\
|\\
\text{CH}_2 \qquad + \text{ DPN (or TPN)} \; \overset{\text{spontaneous}}{\underset{}{\xrightarrow{\text{decomposition}}}} \\
|\\
\text{CH=NH} \\
|\\
\text{COOH} \\
\text{α-Imino} \\
\text{glutamic acid}
\end{array}
\qquad
\begin{array}{l}
\text{COOH} \\
|\\
\text{CH}_2 \\
|\\
\text{CH}_2 + \text{NH}_3 \\
|\\
\text{C=O} \\
|\\
\text{COOH} \\
\text{α-Ketoglutaric acid}
\end{array}
$$

Transamination is the single most important route for the removal of the amino group from amino acids but it is by no means the only one known. The amino group of some L-amino acids may be cleaved by oxidative deamination catalyzed by L-amino acid oxidase with the participation of the flavin-adenine-dinucleotide coenzyme. Here again the corresponding keto acids are formed but ammonia is liberated. Leucine, cysteine, ornithine, and some other amino acids are good substrates for this enzyme but this does not mean that L-amino acid oxidase is of major importance in their decomposition. On the other hand, glycine, serine, threonine, aspartic acid, glutamic acid, ornithine, lysine, and arginine of the L series do not serve as good substrates for this enzyme.

The deamination of some amino acids, such as serine and threonine, may be accomplished by "nonoxidative" means. These reactions are catalyzed by dehydrases.

The decarboxylases, another group of enzymes involved in the intermediary metabolism of amino acid, should also be mentioned. Although decarboxylation is not a major metabolic pathway for the degradation of amino acids, still it carries considerable physiological importance because many compounds formed after the decarboxylation step (histamine, serotonin and so forth) are highly active and important in the function of the mammalian organism. Vitamin B_6 functions as a coenzyme in the decarboxylation reactions.

Glutamic Acid, Aspartic Acid, and Alanine

These three amino acids are transformed to the corresponding keto acids by direct transamination.

$$
\begin{array}{ccc}
\begin{matrix}
CH_3 \\
| \\
CH_2{-}NH_2 \\
| \\
COOH \\
\text{Alanine}
\end{matrix}
&
\xrightarrow[\text{with } \alpha\text{-ketoglutarate}]{\text{transamination}}
&
\begin{matrix}
CH_3 \\
| \\
CO \\
| \\
COOH \\
\text{Pyruvic acid}
\end{matrix}
\end{array}
$$

$$
\begin{array}{ccc}
\begin{matrix}
COOH \\
| \\
CH_2 \\
| \\
CH{-}NH_2 \\
| \\
COOH \\
\text{Aspartic acid}
\end{matrix}
&
\xrightarrow[\text{with } \alpha\text{-ketoglutarate}]{\text{transamination}}
&
\begin{matrix}
COOH \\
| \\
CH_2 \\
| \\
CO \\
| \\
COOH \\
\text{Oxaloacetic acid}
\end{matrix}
\end{array}
$$

$$
\begin{array}{ccc}
\begin{matrix}
COOH \\
| \\
CH_2 \\
| \\
CH_2 \\
| \\
CH{-}NH_2 \\
| \\
COOH \\
\text{Glutamic acid}
\end{matrix}
&
\begin{matrix}
\xrightarrow[]{\text{transamination with oxaloacetate}} \\
\text{or} \\
\text{dehydrogenation} \\
\text{see page 171}
\end{matrix}
&
\begin{matrix}
COOH \\
| \\
CH_2 \\
| \\
CH_2 \\
| \\
CO \\
| \\
COOH \\
\text{Ketoglutaric acid}
\end{matrix}
\end{array}
$$

Arginine, Citrulline, and Ornithine

Arginine and citrulline are transformed to ornithine in the Krebs-Henseleit urea cycle (q.v.). After transamination ornithine is transformed to L-glutamic-γ-semialdehyde. Finally L-glutamic-γ-semialdehyde is enzymatically dehydrogenated to yield glutamic acid, so that the final steps of the degradation of arginine, citrulline, and ornithine are identical with those of the glutamic acid.

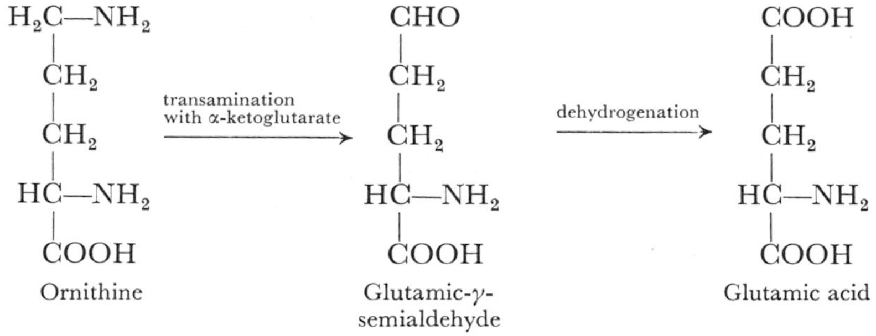

H$_2$C—NH$_2$	CHO	COOH
CH$_2$	CH$_2$	CH$_2$
CH$_2$ $\xrightarrow[\text{with }\alpha\text{-ketoglutarate}]{\text{transamination}}$	CH$_2$ $\xrightarrow{\text{dehydrogenation}}$	CH$_2$
HC—NH$_2$	HC—NH$_2$	HC—NH$_2$
COOH	COOH	COOH
Ornithine	Glutamic-γ-semialdehyde	Glutamic acid

Histidine

The result of the first few steps of histidine degradation is N-formiminoglutamic acid. This is achieved by removal of the α-amino group and hydrolytic cleavage of the imidazole ring. After the enzymatic removal of the N-formimino group, glutamic acid is formed and the final steps of histidine catabolism are those of glutamic acid catabolism.

Since tetrahydrofolic acid is needed to remove the N-formimino group, the appearance of N-formiminoglutamic acid in the urine may disclose folic acid deficiency and is used as a diagnostic test in clinical medicine.

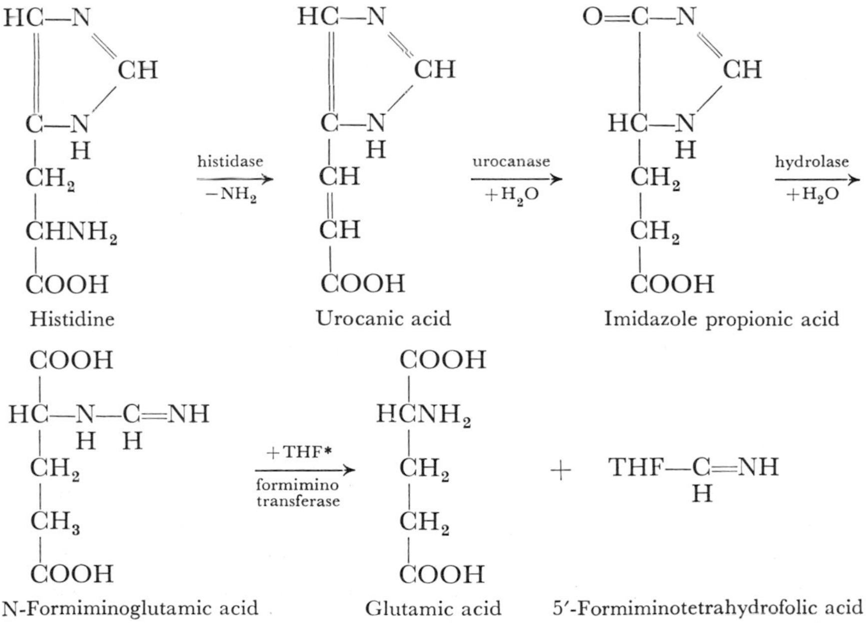

N-Formiminoglutamic acid   Glutamic acid   5'-Formiminotetrahydrofolic acid

---

* Tetrahydrofolic acid.

Another biologically important enzyme, the aromatic L-amino acid decarboxylase, may attack histidine. As the result of this decarboxylation histidine is transformed to histamine.

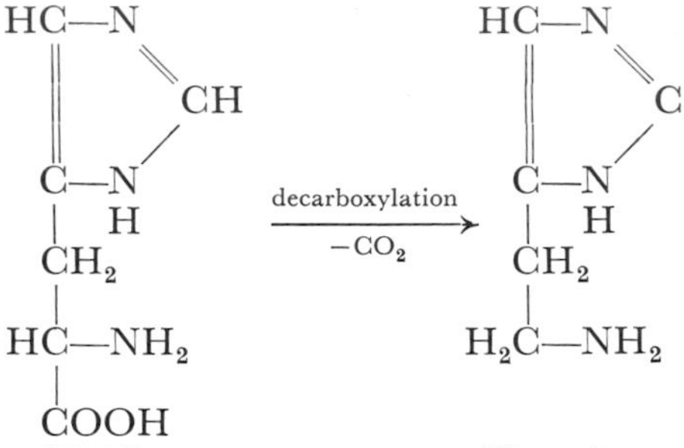

Histidine → (decarboxylation, $-CO_2$) → Histamine

Proline

As the result of oxidative steps and spontaneous ring opening, proline is transformed to glutamic acid.

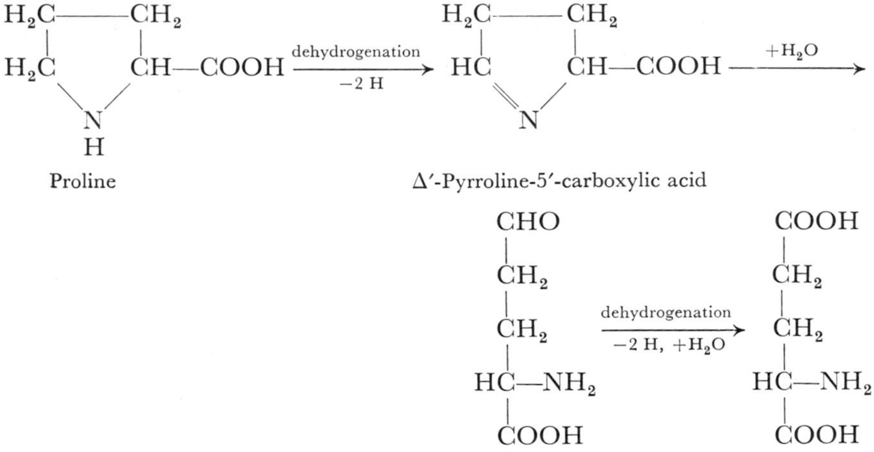

Proline → (dehydrogenation, $-2H$) → Δ'-Pyrroline-5′-carboxylic acid → ($+H_2O$) →

Glutamic-γ-semialdehyde → (dehydrogenation, $-2H$, $+H_2O$) → Glutamic acid

Hydroxyproline

The degradation of hydroxyproline starts with an oxidative step which is followed by a spontaneous ring opening. The γ-hydroxyglutamic semialdehyde so formed is further oxidized and transaminated to yield α-keto-γ-hydroxyglutaric acid. This latter compound undergoes (is subjected to) an aldol fission which produces 1 mole of glyoxylic acid and 1 mole of pyruvic acid.

$$
\begin{array}{c}
\text{H} \\
\text{HO—C———CH}_2 \\
\text{H}_2\text{C} \qquad \text{CH—COOH} \\
\text{N} \\
\text{H}
\end{array}
\xrightarrow[\text{–2 H}]{\text{dehydrogenation}}
$$

Hydroxyproline

$$
\begin{array}{c}
\text{H} \\
\text{HO—C———CH}_2 \\
\text{HC} \qquad \text{CH—COOH} \\
\text{N}
\end{array}
\longrightarrow
\begin{array}{c}
\text{CHO} \\
\text{CHOH} \\
\text{CH}_2 \\
\text{HC—NH}_2 \\
\text{COOH}
\end{array}
\xrightarrow[\text{+H}_2\text{O, –2 H}]{\text{dehydrogenation}}
$$

Δ′-Pyrroline-3-hydroxy-5-carboxylic acid   γ-Hydroxyglutamic semialdehyde

Glyoxylic acid

$$
\begin{array}{c}
\text{COOH} \\
\text{CHOH} \\
\text{CH}_2 \\
\text{HC—NH}_2 \\
\text{COOH}
\end{array}
\xrightarrow[\text{with } \alpha\text{-ketoglutarate}]{\text{transamination}}
\begin{array}{c}
\text{COOH} \\
\text{CHOH} \\
\text{CH}_2 \\
\text{CO} \\
\text{COOH}
\end{array}
\xrightarrow{\text{aldol fission}}
\begin{array}{c}
\text{COOH} \\
\text{CHO} \\
+ \\
\text{CH}_3 \\
\text{CO} \\
\text{COOH}
\end{array}
$$

γ-Hydroxyglutamic acid   α-Keto-γ-hydroxyglutamic acid   Pyruvic acid

Serine, Homoserine, and Threonine

The removal of the amino group from these amino acids is carried out by a nonoxidative deamination. Pyridoxal phosphate is required for the function of the "dehydrase" which is responsible for the deamination (probably common for serine, homoserine, and threonine).

$$
\begin{array}{c}
\text{CH}_2\text{OH} \\
\text{CHNH}_2 \\
\text{COOH}
\end{array}
\xrightarrow[\text{–H}_2\text{O}]{\substack{\text{serine dehydrase} \\ \text{or deaminase}}}
\xrightarrow[\substack{\text{involving} \\ \text{hypothetical} \\ \text{intermediates}}]{\substack{\text{nonenzymatic} \\ \text{rearrangements}}}
\begin{array}{c}
\text{CH}_3 \\
\text{CO} \\
\text{COOH}
\end{array}
+ \text{NH}_3
$$

L-Serine   Pyruvic acid

$$
\begin{array}{c}
\text{CH}_2\text{OH} \\
\text{CH}_2 \\
\text{CHNH}_2 \\
\text{COOH}
\end{array}
\xrightarrow{\substack{\text{homoserine} \\ \text{dehydrase or} \\ \text{deaminase}}}
\xrightarrow[\substack{\text{involving} \\ \text{hypothetical} \\ \text{intermediates}}]{\substack{\text{nonenzymatic} \\ \text{rearrangement}}}
\begin{array}{c}
\text{CH}_3 \\
\text{CH}_2 \\
\text{CO} \\
\text{COOH}
\end{array}
+ \text{NH}_3
$$

p-Homoserine   α-Ketobutyric acid

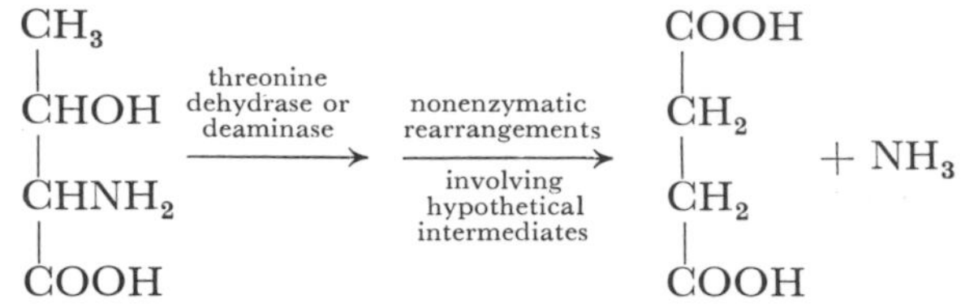

$$
\begin{array}{ccc}
CH_3 & & COOH \\
| & & | \\
CHOH & \xrightarrow[\text{deaminase}]{\text{threonine}} \xrightarrow[\substack{\text{involving}\\\text{hypothetical}\\\text{intermediates}}]{\text{nonenzymatic}} & CH_2 \\
| & \text{dehydrase or} \quad \text{rearrangements} & | \quad + NH_3 \\
CHNH_2 & & CH_2 \\
| & & | \\
COOH & & COOH \\
\text{Threonine} & & \text{Succinic acid}
\end{array}
$$

Leucine, Isoleucine, and Valine

The initial steps of the catabolism of these three amino acids are the same. The first step is transamination, yielding the corresponding keto acids. The next step involves activation with coenzyme A and oxidative decarboxylation. This step bears some resemblance to the oxidative decarboxylation of the fatty acids. Probably the same enzyme is responsible for the oxidative decarboxylation of these three amino acids. The third step is dehydrogenation, leading to the formation of unsaturated compounds. There are individual differences in the further steps involved in the degradation of leucine, isoleucine, and valine, but with all three there is some similarity to the oxidation of fatty acids.

The individual steps of the degradation of leucine are shown here. With respect to isoleucine and valine, we merely mention that their decomposition ultimately yields propionic acid (CH_3—CH_2—$COOH$).

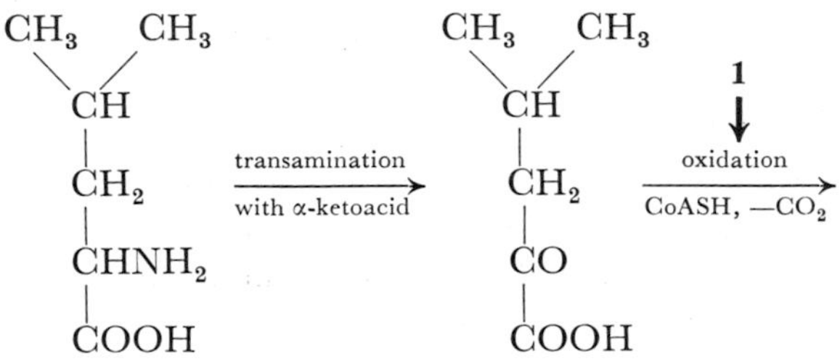

$$
\begin{array}{cccc}
CH_3 \diagdown \diagup CH_3 & & CH_3 \diagdown \diagup CH_3 & \mathbf{1} \\
CH & & CH & \downarrow \\
| & \xrightarrow[\text{with } \alpha\text{-ketoacid}]{\text{transamination}} & | & \xrightarrow[\text{CoASH, } -CO_2]{\text{oxidation}} \\
CH_2 & & CH_2 & \\
| & & | & \\
CHNH_2 & & CO & \\
| & & | & \\
COOH & & COOH & \\
\text{L-Leucine} & & \alpha\text{-Ketoisocaproic acid} &
\end{array}
$$

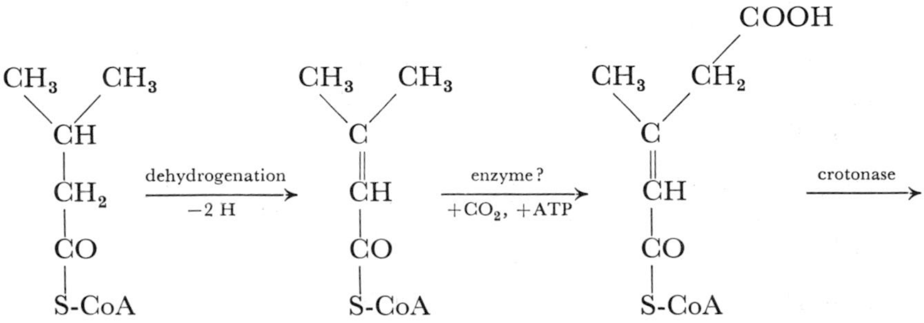

$$
\begin{array}{cccccc}
& & & & & COOH \\
CH_3 \diagdown \diagup CH_3 & & CH_3 \diagdown \diagup CH_3 & & CH_3 \diagdown \diagup CH_2 & \\
CH & & C & & C & \\
| & \xrightarrow[-2H]{\text{dehydrogenation}} & \| & \xrightarrow[+CO_2, +ATP]{\text{enzyme?}} & \| & \xrightarrow{\text{crotonase}} \\
CH_2 & & CH & & CH & \\
| & & | & & | & \\
CO & & CO & & CO & \\
| & & | & & | & \\
\text{S-CoA} & & \text{S-CoA} & & \text{S-CoA} & \\
\text{Isovaleryl-CoA} & & \beta\text{-Methyl-crotonyl-CoA} & & \beta\text{-Methyl-glutaconyl-CoA} &
\end{array}
$$

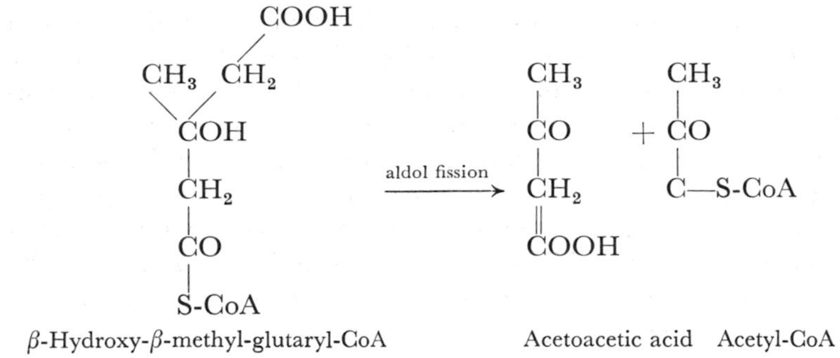

$$
\begin{array}{c}
COOH \\
\diagup \\
CH_3 \quad CH_2 \\
\diagdown \diagup \\
COH \\
| \\
CH_2 \\
| \\
CO \\
| \\
S\text{-}CoA
\end{array}
\quad
\xrightarrow{\text{aldol fission}}
\quad
\begin{array}{c}
CH_3 \\
| \\
CO \\
| \\
CH_2 \\
\| \\
COOH
\end{array}
\;+\;
\begin{array}{c}
CH_3 \\
| \\
CO \\
| \\
C\text{---}S\text{-}CoA
\end{array}
$$

β-Hydroxy-β-methyl-glutaryl-CoA   Acetoacetic acid   Acetyl-CoA

Cystine and Cysteine

Cystine and cysteine have a common catabolic pathway since cystine, after reduction, yields two molecules of cysteine. The main pathway for the degradation of cysteine starts with oxidative steps leading to the formation of cysteic acid. Cysteic acid is transformed into β-sulfonylpyruvic acid by transamination. β-sulfonylpyruvic acid is spontaneously degraded to pyruvic acid and sulfate.

If cysteic acid is decarboxylated rather than transaminated, taurine is formed. The physiological importance of taurine lies in the fact that it is a part of the conjugated bile acids (taurocholic acid).

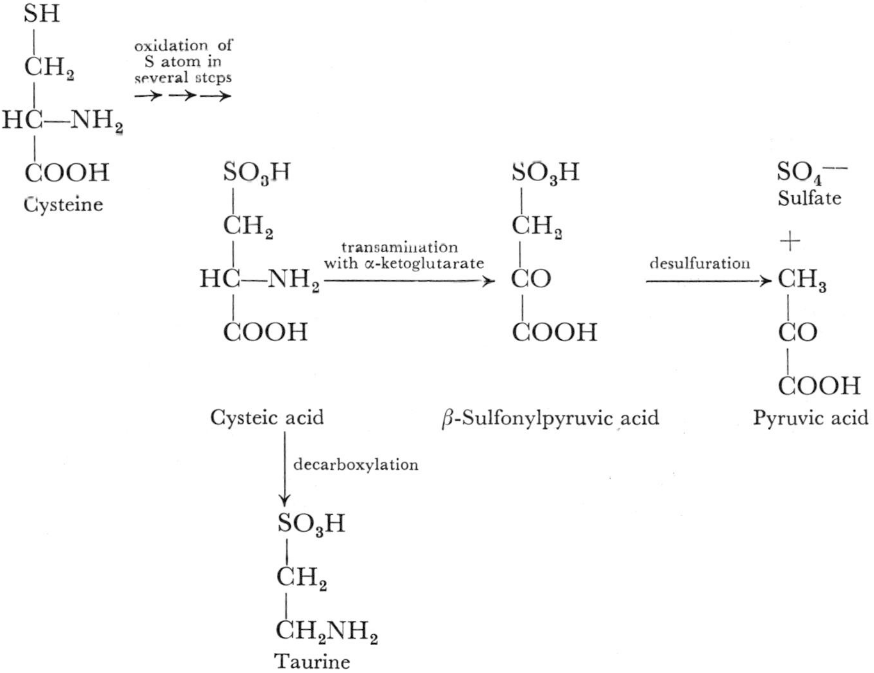

$$
\begin{array}{c}
SH \\
| \\
CH_2 \\
| \\
HC\text{---}NH_2 \\
| \\
COOH
\end{array}
\xrightarrow[\text{several steps}]{\substack{\text{oxidation of}\\ \text{S atom in}}}
\begin{array}{c}
SO_3H \\
| \\
CH_2 \\
| \\
HC\text{---}NH_2 \\
| \\
COOH
\end{array}
\xrightarrow[\text{with } \alpha\text{-ketoglutarate}]{\text{transamination}}
\begin{array}{c}
SO_3H \\
| \\
CH_2 \\
| \\
CO \\
| \\
COOH
\end{array}
\xrightarrow{\text{desulfuration}}
\begin{array}{c}
SO_4^{--} \\
\text{Sulfate} \\
+ \\
CH_3 \\
| \\
CO \\
| \\
COOH
\end{array}
$$

Cysteine   Cysteic acid β-Sulfonylpyruvic acid   Pyruvic acid

$$
\downarrow \text{decarboxylation}
$$

$$
\begin{array}{c}
SO_3H \\
| \\
CH_2 \\
| \\
CH_2NH_2
\end{array}
$$

Taurine

Methionine

Methionine is transformed to cysteine first and its further catabolic fate is similar to that of cysteine.

The methionine-cysteine transformation involves the participation of ATP and leads to the formation of a physiologically important intermediate called S-adenosylmethionine. This compound is also known as active methionine, and it plays a key role in transmethylation reactions. S-Adenosylmethionine is transformed to homocysteine in two steps. First the S-methyl group is transferred to a suitable acceptor; then the S-adenosyl group is detached from the rest of the molecule. Homocysteine combines with a serine molecule to form cystathionine. When cystathionine is cleaved, the terminal CH_2 group of homocysteine remains on the serine molecule and the products are cysteine and homoserine.* Further degradation of methionine is similar to that of cysteine.

$$
\begin{array}{ccc}
CH_3 & & CH_3 \\
| & & | \\
S & & S-CH_2-Adenosine \\
| & & | \\
CH_2 & & CH_2 \\
| & & | \\
CH_2 & \xrightarrow[\substack{-P_i,\ -\text{pyrophosphate}}]{\substack{\text{methionine-}\\ \text{activating}\\ \text{enzyme}\\ +ATP}} & CH_2 \\
| & & | \\
HC-NH_2 & & HC-NH_2 \\
| & & | \\
COOH & & COOH \\
\text{Methionine} & & \text{S-Adenosylmethionine}
\end{array}
$$

$$\xrightarrow[\substack{\text{methyl group}\\ \text{acceptor}}]{\substack{\text{methyl group}\\ \text{transfer enzyme}}}$$

$$
\begin{array}{cc}
S-CH_2-Adenosine & \\
| & \\
CH_2 & SH \\
| & | \\
CH_2 & CH_2 \\
| & | \\
HC-NH_2 & HC-NH_2 \\
| & | \\
COOH & COOH \\
\text{S-Adenosylhomocysteine} & \text{Homocysteine}
\end{array}
$$

(between them: $\xrightarrow[-\text{adenosine}]{+H_2O}$)

1
↓ cystathionine synthetase + serine, $-H_2O$

$$
\begin{array}{ccc}
H_2C-S-CH_2 & SH & CH_2OH \\
| \quad\quad | & | & | \\
HC-NH_2 \quad CH_2 & CH_2 & CH_2 \\
| \quad\quad | & | & | \\
COOH \quad HC-NH_2 & HC-NH_2 + HC-NH_2 \\
| & | & | \\
COOH & COOH & COOH \\
\text{Cystathionine} & \text{Cystheine} & \text{Homoserine}
\end{array}
$$

(arrow: $\xrightarrow{\substack{\text{cystathionase}\\ \text{(hydrolytic cleavage)}}}$)

$$
\begin{array}{c}
CH_3 \\
| \\
CH_2 \\
| \\
CO \\
| \\
COOH \\
\text{α-Ketobutyric acid}
\end{array}
$$

(*deamination*)

---

* Homoserine is degraded to α-ketobutyric acid by deamination.

Phenylalanine and Tyrosine

The most important catabolic pathway of both these amino acids is the same. First phenylalanine is hydroxylated to tyrosine, and then tyrosine participates in a transamination reaction with ketoglutaric acid. Further enzymatic steps result in the oxidative cleavage of the aromatic ring and fumarylacetoacetic acid is formed. This latter compound undergoes a hydrolytic fission to yield 1 mole of fumaric acid and 1 mole of acetoacetic acid.

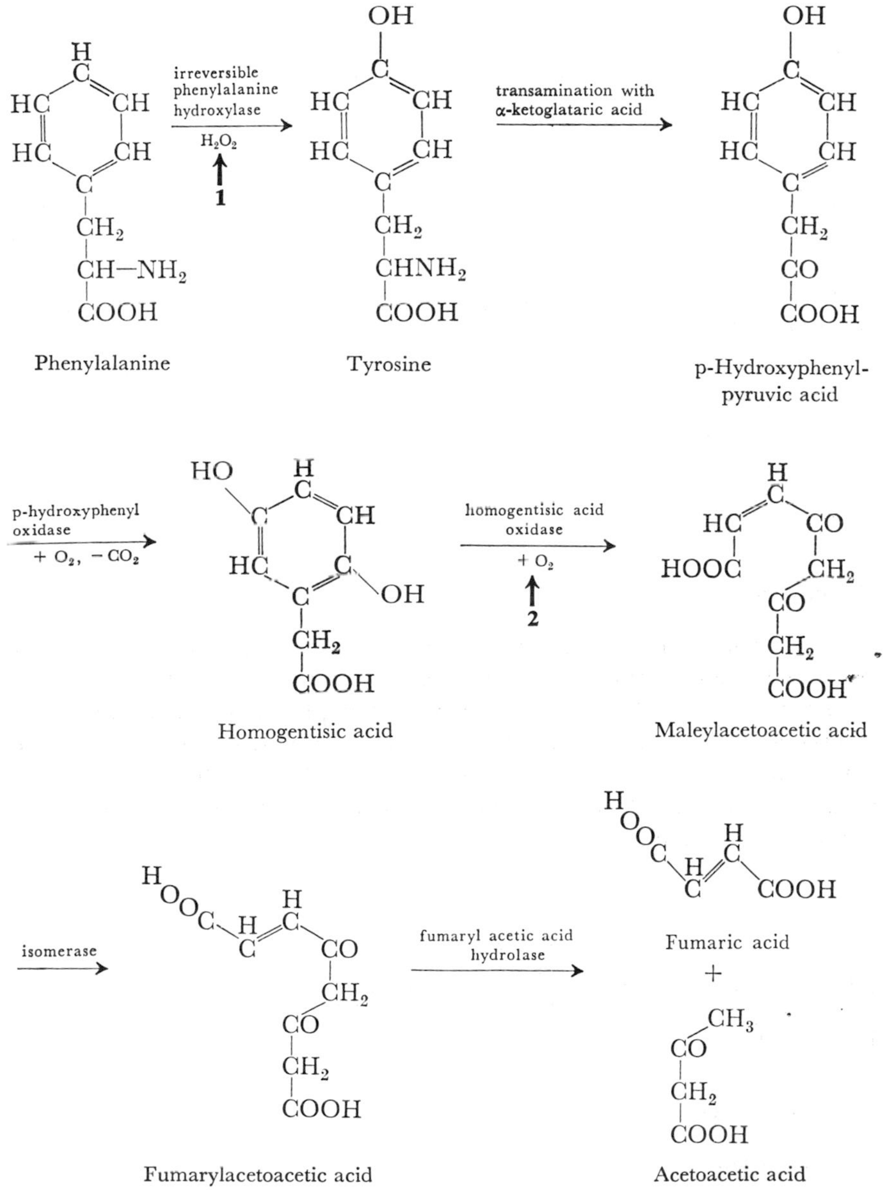

Phenylalanine   Tyrosine   p-Hydroxyphenyl-pyruvic acid

Homogentisic acid   Maleylacetoacetic acid

Fumarylacetoacetic acid   Fumaric acid + Acetoacetic acid

Tyrosine may be decarboxylated by the aromatic L-amino acid de-carboxylase to form tyramine. The same amino acid may also be oxidized by tyrosinase to dihydroxyphenylalanine and melanin. Dihydroxyphenyl-alanine, after some enzymatic transformations, leads to the formation of norepinephrine and epinephrine.

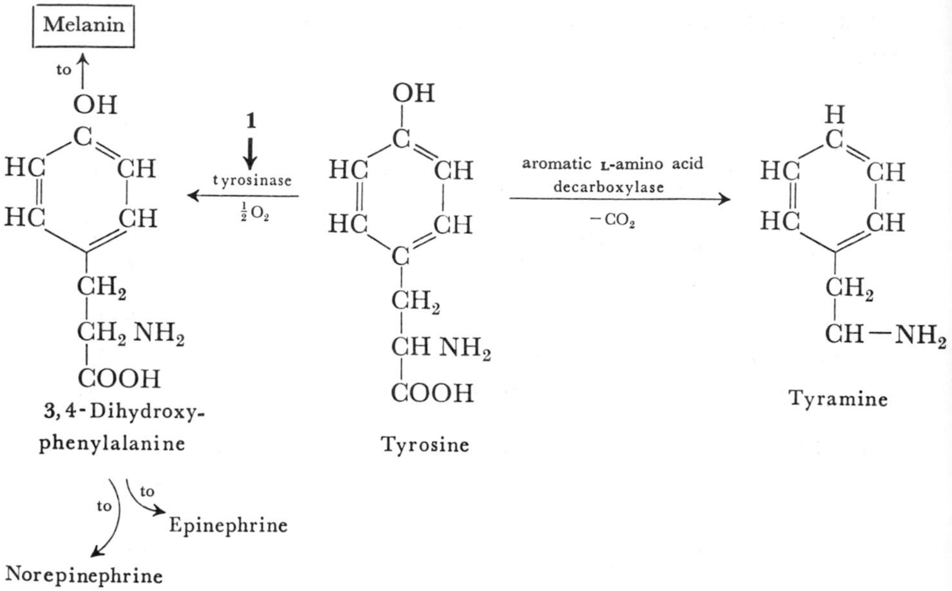

3,4-Dihydroxy-phenylalanine

Tyrosine

Tyramine

Norepinephrine

Epinephrine

Tryptophan

The catabolism of tryptophan follows two main pathways. The first involves a hydroxylation step on the indole ring and the decarboxylation of the alanine residue. These steps transform tryptophan to serotonin (hydroxy-tryptamine). This compound has a strong vasoconstrictor effect in the mammalian organism. Serotonin is further oxidized to S-hydroxyindole-acetic acid, which is the end product of this pathway and is eliminated from the body by the kidneys.

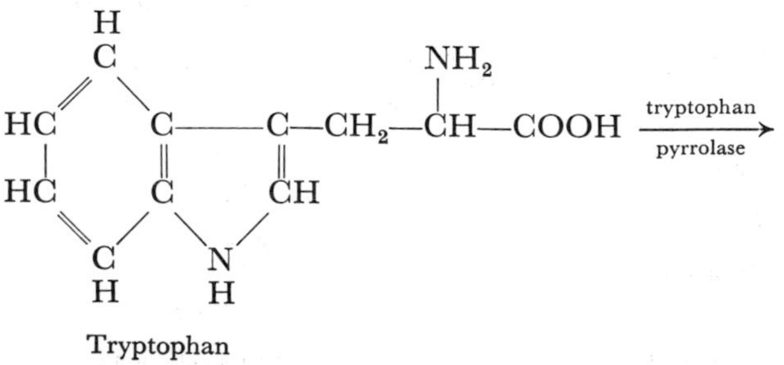

Tryptophan

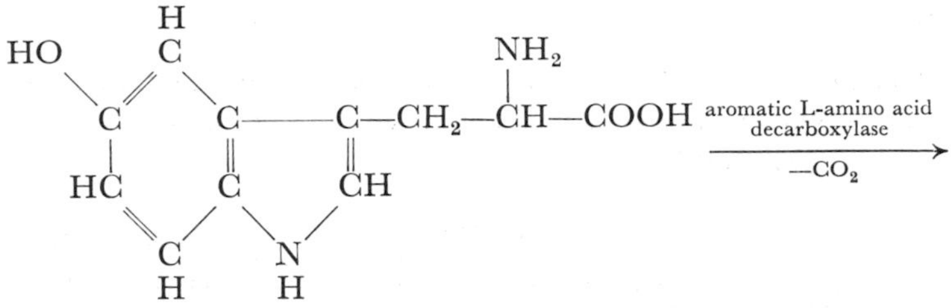

5′-Hydroxytryptophan

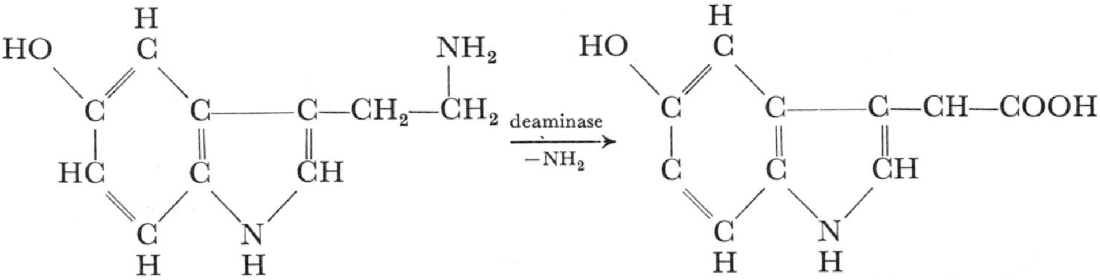

$\xrightarrow[-CO_2]{\text{aromatic L-amino acid decarboxylase}}$

Serotonin

$\xrightarrow[-NH_2]{\text{deaminase}}$

5′-Hydroxyindoleacetic acid

The second catabolic pathway for tryptophan is very complicated. We merely wish to mention that at one point kynurenine is formed as an intermediate. For the further degradation of kynurenine the participation of vitamin B_6 is necessary. If the mammalian organism is deficient in vitamin B_6 kynurenine is transformed to xanthurenic acid, which appears in the urine and may be used as an indication for possible vitamin B_6 deficiency. The end result of the further physiological degradation of kynurenine is 1 mole of fumaric acid and 1 mole of acetoacetic acid per mole of tryptophan.

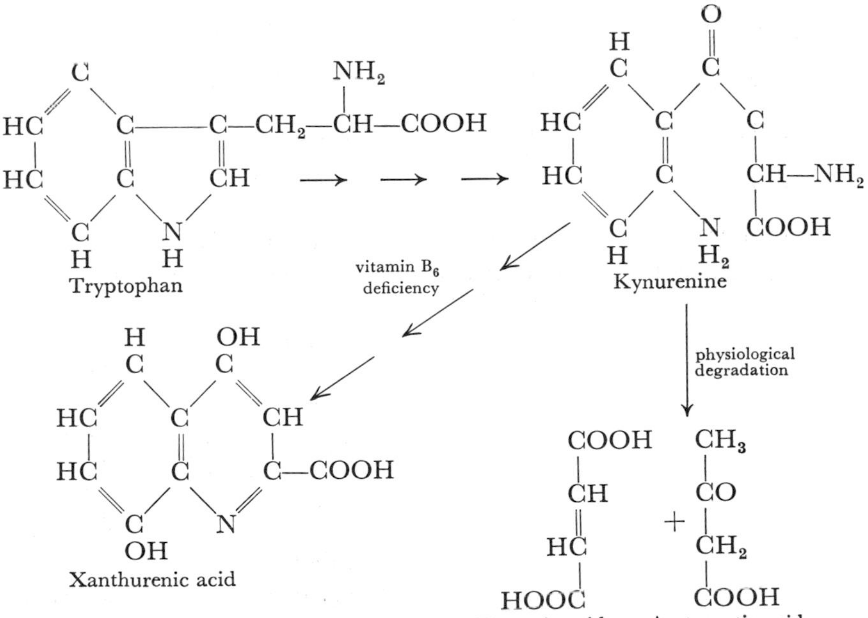

Tryptophan

$\longrightarrow \longrightarrow \longrightarrow$

Kynurenine

vitamin B_6 deficiency

physiological degradation

Xanthurenic acid

Fumaric acid   Acetoacetic acid

Glycine

Among the several pathways known for the degradation of this metabolically very active amino acid the nature of the main catabolic pathway is not conclusively established.

a

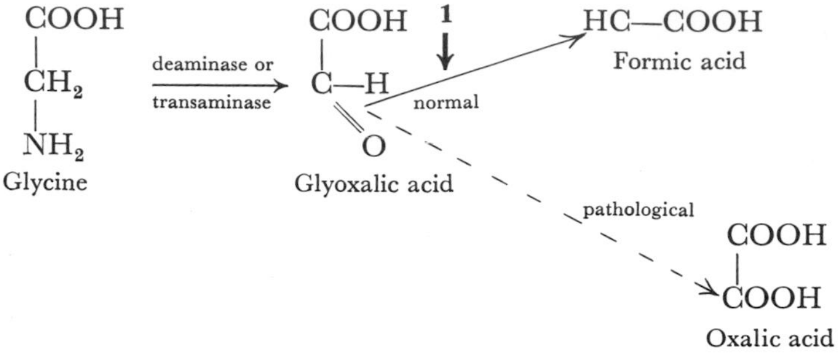

$$
\begin{array}{ccc}
\text{COOH} & & \text{COOH} \quad \mathbf{1} \\
| & \xrightarrow[\text{transaminase}]{\text{deaminase or}} & | \quad \downarrow \\
\text{CH}_2 & & \text{C—H} \\
| & & \parallel \quad \text{normal} \\
\text{NH}_2 & & \text{O} \\
\text{Glycine} & & \text{Glyoxalic acid}
\end{array}
$$

HC—COOH
Formic acid

pathological

COOH
|
COOH
Oxalic acid

b

$$
\begin{array}{c}
\text{COOH} \\
| \\
\text{CH}_2 \\
| \\
\text{NH}_2 \\
\text{Glycine}
\end{array}
+
\begin{array}{c}
\text{COSR} \\
| \\
\text{CH}_2 \\
| \\
\text{CH}_2 \\
| \\
\text{COOH} \\
\text{Succinyl-CoA}
\end{array}
\xrightarrow{\text{synthetase}}
\begin{array}{c}
\text{COOH} \\
| \\
\text{HC—NH}_2 \\
| \\
\text{CO} \\
| \\
\text{CH}_2 \\
| \\
\text{CH}_2 \\
| \\
\text{COOH} \\
\alpha\text{-Amino-}\beta\text{-ketoadipic acid}
\end{array}
\xrightarrow[-CO_2]{\substack{\text{nonenzymatic} \\ \text{decarboxylation}}}
$$

$$
\begin{array}{c}
\text{H}_2\text{C—NH}_2 \\
| \\
\text{CO} \\
| \\
\text{CH}_2 \\
| \\
\text{CH}_2 \\
| \\
\text{COOH} \\
\delta\text{-Aminolevulinic acid}
\end{array}
\longrightarrow
\begin{array}{c}
\text{CHO} \\
| \\
\text{CO} \\
| \\
\text{CH}_2 \\
| \\
\text{CH}_2 \\
| \\
\text{COOH} \\
\alpha\text{-Ketoglutaric} \\
\text{semialdehyde}
\end{array}
\longrightarrow
\begin{array}{c}
\text{COOH} \\
| \\
\text{HCOH} \\
| \\
\text{CH}_2 \\
| \\
\text{CH}_2 \\
| \\
\text{COOH} \\
\alpha\text{-Hydroxyglutaric acid}
\end{array}
\longrightarrow
$$

$$
\begin{array}{c}
\text{COOH} \\
| \\
\text{CO} \\
| \\
\text{CH}_2 \\
| \\
\text{CH}_2 \\
| \\
\text{COOH} \\
\alpha\text{-Ketoglutaric acid}
\end{array}
\xrightarrow[-2\text{ H, }-CO_2]{+\text{CoA}}
\begin{array}{c}
\text{COSR} \\
| \\
\text{CH}_2 \\
| \\
\text{CH}_2 \\
| \\
\text{COOH} \\
\text{Succinyl-CoA}
\end{array}
$$

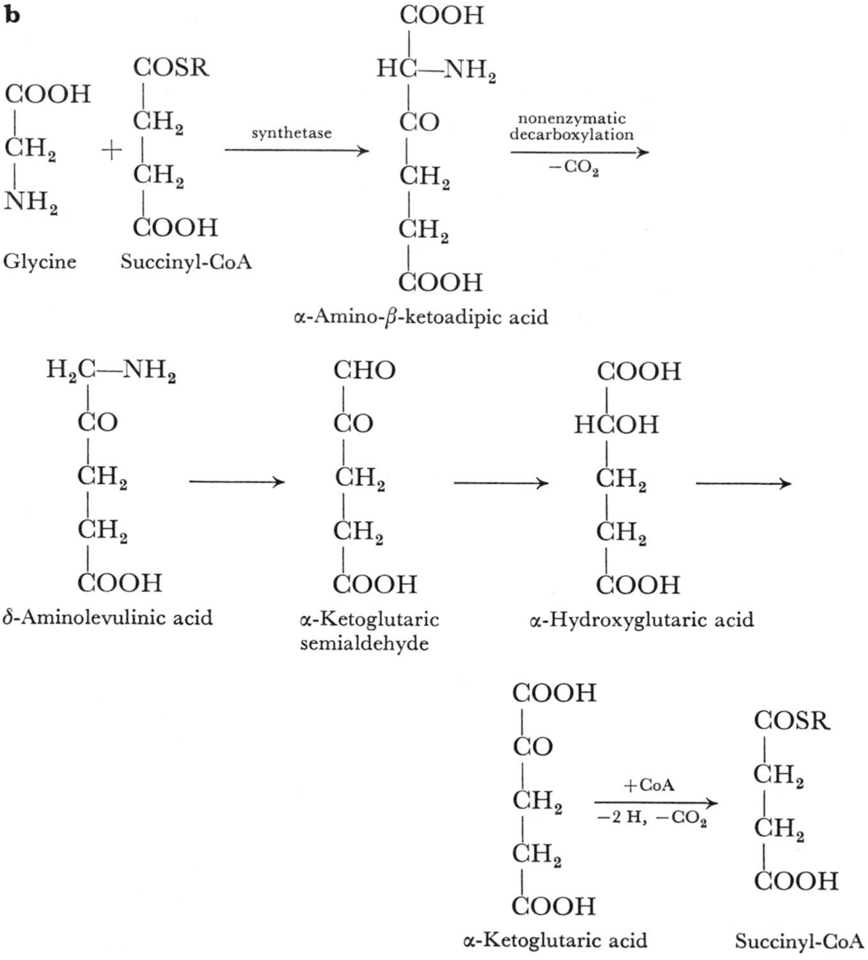

Glycine is completely oxidized in this pathway. The succinyl-coenzyme A used in the first synthetic step is recovered in the last step.

Glycine may be incorporated into the following physiologically important compounds: (1) porphyrin ring of hemin; (2) purine and pyrimidine bases of nucleic acids (Chapter 10); (3) choline,

$$CH_3-\underset{\overset{|}{C}H_3}{\overset{CH_3}{N}}-CH_2-CH_2OH;$$

Choline

(4) glycogen (via Krebs cycle; see end products of pathway a); (5) fatty acids (see pathway b); (6) creatine; and (7) glutathione.

Glycine may give rise to a one-carbon compound by producing formic acid (pathway a), and it also may be used by the mammalian organism for conjugation reactions. Conjugation with cholic acid leads to glycocholic acid, and with benzoic acid it yields hippuric acid. These reactions, performed by the liver, serve detoxication purposes.

UREA FORMATION

During the degradation process of various amino acids ammonia is being formed constantly. Under normal conditions the level of ammonia in blood and urine is very low and does not correspond to the amount of ammonia set free by the catabolism of amino acids. It has been established that the ammonia produced by the degradation of protein is transformed into urea and excreted from the body as such. Krebs and Henseleit discovered that the synthesis of urea takes place in the liver. They established that ornithine, citrulline, and arginine participate in the urea formation without being consumed during this process. The synthesis of urea via ornithine is known as the Krebs-Henseleit cycle.

First ammonia is trapped in the form of carbamyl phosphate. This is achieved by the carbamyl phosphate synthetase in the presence of CO_2 and ATP. It is the carbamyl groups of the carbamyl phosphate which is transferred to ornithine by a transcarbamylase enzyme. At this point the ammonia makes its entry into the ornithine cycle while during the same reaction ornithine is transformed to citrulline. This latter compound in the presence of a synthetase enzyme interacts with aspartic acid and leads to the formation of argininosuccinate. In the next step an enzymatic cleavage produces arginine and fumaric acid and finally arginase splits arginine to urea and ornithine.

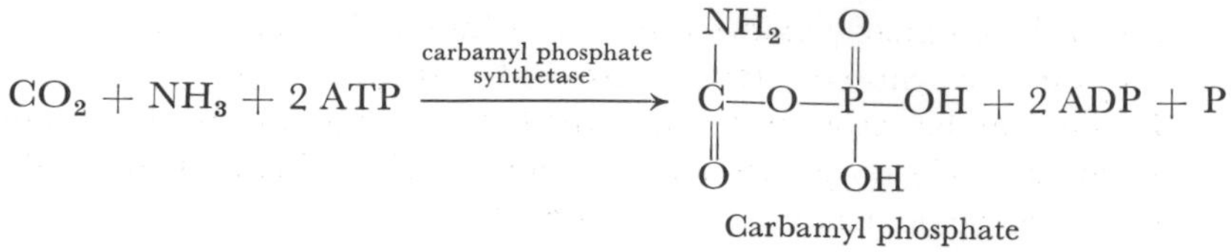

carbamyl phosphate synthetase
Carbamyl phosphate

1
ornithine transcarbamylase
Carbamyl phosphate
Ornithine
Citrulline

2
Argininosuccinate synthetase
Citrulline
Aspartic acid
Argininosuccinate

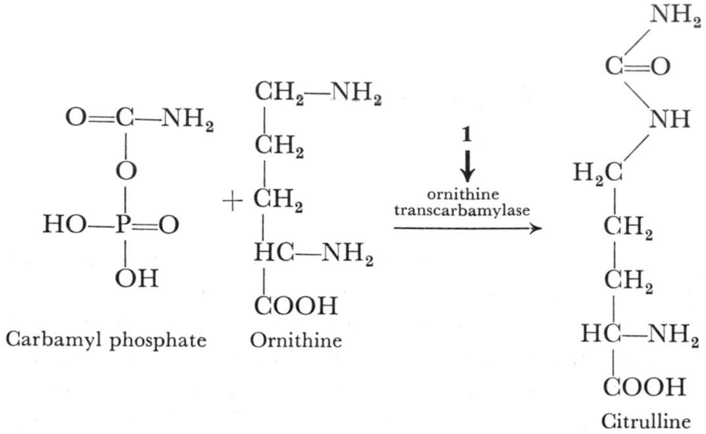

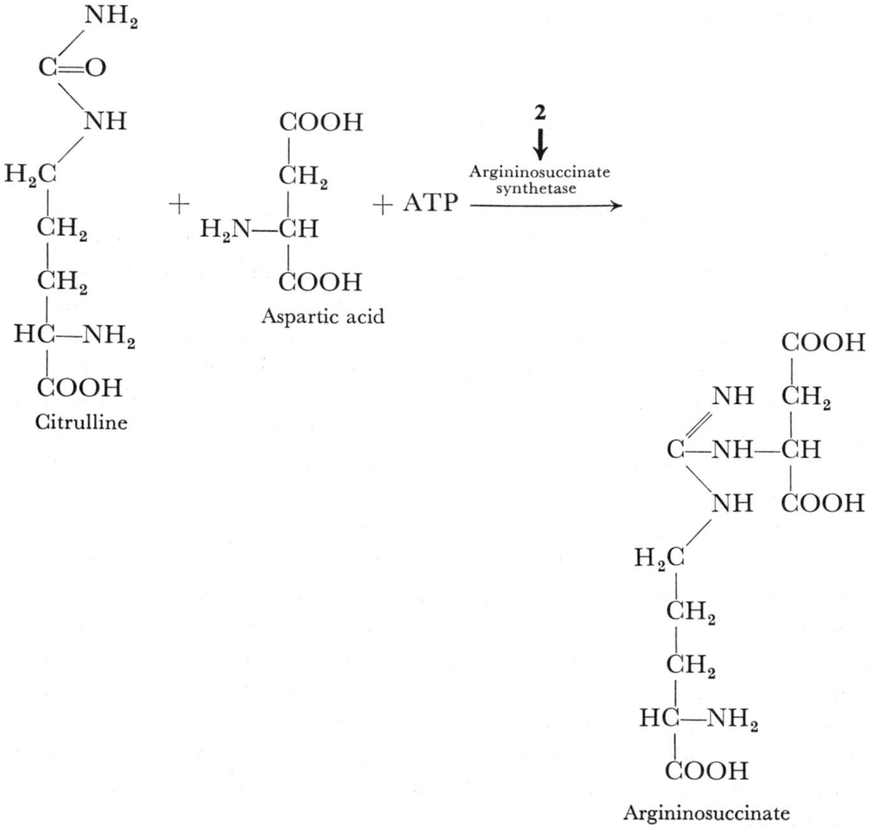

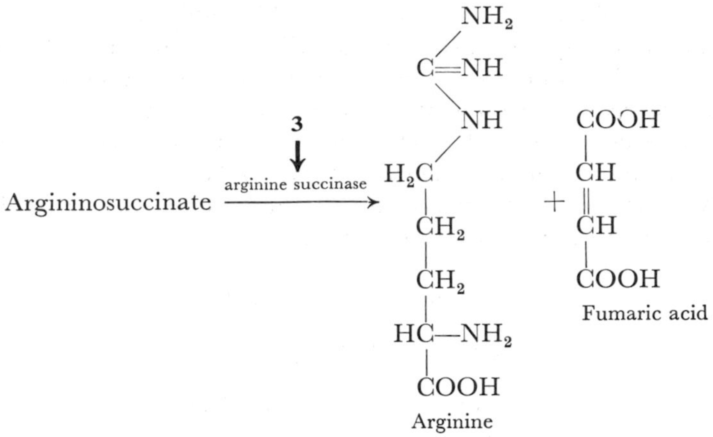

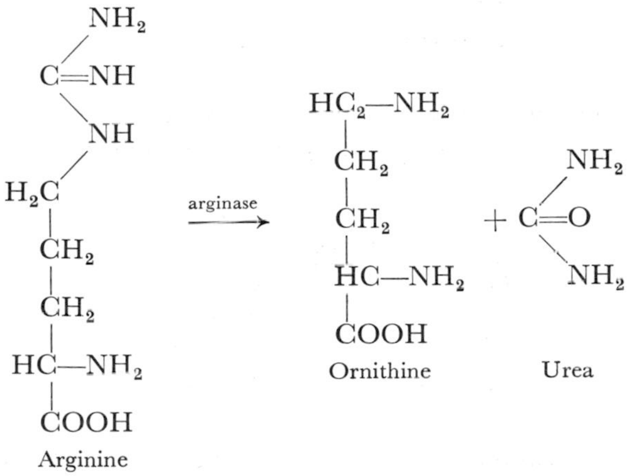

$$
\text{Argininosuccinate} \xrightarrow[\text{arginine succinase}]{\textbf{3}}
\begin{array}{c}
\text{NH}_2 \\
\diagup \\
\text{C}=\text{NH} \\
\diagdown \\
\text{NH} \\
\diagup \\
\text{H}_2\text{C} \\
| \\
\text{CH}_2 \\
| \\
\text{CH}_2 \\
| \\
\text{HC--NH}_2 \\
| \\
\text{COOH} \\
\text{Arginine}
\end{array}
\quad + \quad
\begin{array}{c}
\text{COOH} \\
| \\
\text{CH} \\
\| \\
\text{CH} \\
| \\
\text{COOH} \\
\text{Fumaric acid}
\end{array}
$$

$$
\begin{array}{c}
\text{NH}_2 \\
\diagup \\
\text{C}=\text{NH} \\
\diagdown \\
\text{NH} \\
\diagup \\
\text{H}_2\text{C} \\
| \\
\text{CH}_2 \\
| \\
\text{CH}_2 \\
| \\
\text{HC--NH}_2 \\
| \\
\text{COOH} \\
\text{Arginine}
\end{array}
\xrightarrow{\text{arginase}}
\begin{array}{c}
\text{HC}_2\text{--NH}_2 \\
| \\
\text{CH}_2 \\
| \\
\text{CH}_2 \\
| \\
\text{HC--NH}_2 \\
| \\
\text{COOH} \\
\text{Ornithine}
\end{array}
\quad + \quad
\begin{array}{c}
\text{NH}_2 \\
\diagup \\
\text{C}=\text{O} \\
\diagdown \\
\text{NH}_2 \\
\text{Urea}
\end{array}
$$

BIOSYNTHESIS OF THE AMINO ACIDS

The mammalian organism is able to synthesize nine amino acids using metabolic products originating from nutrients other than proteins. The synthesis of the other amino acids is either too slow or it is not carried out at all. The following amino acids seem to fall in this latter category: leucine, isoleucine, valine, methionine, lysine, threonine, phenylalanine, tryptophan, histidine, and arginine. These amino acids are called "essential" because they have to be supplied to the mammalian organism to ensure its proper functioning.

The nonessential amino acids may originate by synthesis from carbohydrates, fats, or the essential amino acids. Most of these synthetic processes involve the incorporation of ammonia into organic acids to form the amino nitrogen. In this process glutamic dehydrogenase plays the central role. The liberation of NH_3 and ketoglutarate from glutamic acid is readily reversible. Therefore the addition of ammonia to α-ketoglutarate may lead to the formation of glutamic acid. Glutamic acid readily participates in trans-aminations (mainly in the liver and kidneys) and serves as an amino donor for other keto acids, transforming them into amino acids. Some of the biosyn-thetic pathways of several nonessential amino acids are described here.

GLUTAMIC ACID. α-Ketoglutarate is an intermediate in the Krebs cycle and can be aminated immediately by ammonia.

ASPARTIC ACID. Oxaloacetate, an intermediate in the Krebs cycle, after transaminations yields aspartic acid.

$$
\begin{array}{ccc}
\text{COOH} & & \text{COOH} \\
| & & | \\
\text{CH}_2 & \xrightarrow[\text{transamination}]{\text{glutamic}} & \text{CH}_2 \\
| & & | \\
\text{CO} & & \text{CHNH}_2 \\
| & & | \\
\text{COOH} & & \text{COOH} \\
\text{Oxaloacetic acid} & & \text{Aspartic acid}
\end{array}
$$

ALANINE. Pyruvate is formed in glycolysis and in the oxidation of lactate. Pyruvate after transamination gives rise to alanine.

$$
\begin{array}{ccc}
\text{CH}_3 & & \text{CH}_3 \\
| & & | \\
\text{CO} & \xrightarrow[\text{transamination}]{\text{glutamic}} & \text{CHNH}_2 \\
| & & | \\
\text{COOH} & & \text{COOH} \\
\text{Pyruvate} & & \text{Alanine}
\end{array}
$$

PROLINE. Biosynthesis of proline proceeds in the reverse direction but involves the same intermediates as its catabolism. The enzymes involved in the synthesis, however, are probably different from those participating in the catabolism.

HYDROXYPROLINE. Hydroxyproline is synthesized from proline by hydroxylation. It is interesting to note that hydroxylation of the proline moiety does not take place until after proline is incorporated into collagen.

SERINE. Mammalian tissues can synthesize serine in different ways. Glycerol can be dehydrogenated and transaminated (with alanine) to yield serine. Phosphoglycerol may follow a similar course and yield phos-phoserine, which after losing its phosphate group also produces serine.

$$
\begin{array}{ccccccc}
\text{CH}_2\text{OH} & & \text{CH}_2\text{OH} & & \text{CH}_2\text{OH} & & \text{CH}_2\text{OH} \\
| & \xrightarrow[-2\text{H}]{\text{dehydrogenase}} & | & \xrightarrow[-2\text{H}]{\text{dehydrogenase}} & | & \xrightarrow[\text{(with valine)}]{\text{transaminase}} & | \\
\text{CHOH} & & \text{CHOH} & & \text{CO} & & \text{CHNH}_2 \\
| & & | & & | & & | \\
\text{CH}_2\text{OH} & & \text{COOH} & & \text{COOH} & & \text{COOH} \\
\text{Glycerol} & & \text{Glyceric acid} & & \text{Hydroxypyruvate} & & \text{Serine}
\end{array}
$$

Serine is formed from glycine by addition of a hydroxymethyl group. The source for the transferred hydroxymethyl group is sarcosine, and tetrahydrofolic acid plays a key role in the transfer.

$$
\begin{array}{cccc}
\text{COOH} & & \text{COOH} & \\
| & & | & \\
\text{CH}_2 & + \text{THF—CH}_2\text{OH} \rightleftharpoons & \text{CHNH}_2 & + \text{THFH} \\
| & & | & \\
\text{NH}_2 & & \text{CH}_2\text{OH} & \\
\text{Glycine} & \text{Hydroxymethyl-tetra-} & \text{Serine} & \text{Tetrahydrofolic acid} \\
& \text{hydrofolic acid} & &
\end{array}
$$

GLYCINE. Glycine may be derived from serine as shown above; thus ultimately the carbon atoms of glycine and serine may come from the same sources.

TYROSINE. Tyrosine may be synthesised by the hydroxylation of phenylalanine (see catabolism of phenylalanine and tyrosine).

CYSTEINE AND CYSTINE. The synthesis of cysteine and cystine involves the same steps as their degradation. Essentially, the carbon skeleton is provided by serine and the sulfur by methionine.

ORNITHINE AND CITRULLINE. Ornithine and citrulline may be obtained from arginine as discussed in the Krebs-Henseleit urea cycle.

SUMMARY OF THE DISORDERS OF AMINO ACID METABOLISM

Garrod in 1908 suggested that homogentisic acid that appears in the urine of some sick children is an intermediate of the catabolism of phenylalanine and tyrosine. He connected the occurrence of this intermediate with the lack of a specific enzyme needed for the further degradation of homogentisic acid. This disease (alkaptonuria) and some others that Garrod observed—cystinuria, pentosuria, and albinism—make their appearance in children. Their occurrence is familial and they are often observed in families in which consanguineous marriages have taken place. Garrod called these diseases "inborn errors of metabolism." The rapidly advancing analytical techniques and the increasing biochemical understanding of amino acid metabolism have made it possible to prove the existence of a large number of these disorders.

A wide variety of symptoms are usually connected with the appearance of large amounts of a specific amino acid, or several amino acids, and some of their metabolic products in blood and urine. The causal connection between the metabolic disorder and the clinical symptoms is often unknown. The symptoms may be caused by the accumulation of the given amino acid or its metabolic product or by the lack of formation of certain biologically important products. It must be emphasized that not all aminoaciduria is caused by an anomaly of the metabolism. Often the cause is simply a lowered capacity of the kidneys to reabsorb amino acids.

We will briefly survey some disorders of the amino acid metabolism.

HYPERGLYCINEMIA (GLYCINEMIA).   In this disorder there is a block in glycine-serine conversion.

HYPEROXALURIA.   This is caused by a block in the degradation of glycine so that glyoxalic acid cannot be converted to formic acid (see arrow 1, page 182).   The increased amount of oxalate in the urinary system may lead to the formation of stones and eventually to renal failure.

MAPLE SYRUP DISEASE.   The name reflects the characteristic odor of urine excreted by those with this disease.   The metabolic block in the catabolism of leucine, isoleucine, and valine occurs right after the transamination (see arrow 1, page 176).   Therefore the corresponding keto acids and the amino acids involved accumulate in the tissues, blood, and urine and are responsible for the characteristic odor of the urine.

HOMOCYSTINURIA.   The enzyme missing is cystathionine synthetase (see arrow 1, page 178), and the conversion of serine plus methionine to cysteine and homoserine is blocked.   Since cystathionine is normally present in high concentration in the brain tissue, the failure of its formation has been implicated as the etiologic factor of the symptoms of this disease.

HYPERAMMONEMIA.   The accumulation of ammonia in tissues, blood, and urine appears to be a common symptom in several disorders involving the enzymes of the urea cycle.   In this group of disorders of amino acid metabolism the enzymes needed for the conversion of ornithine to citrulline (arrow 1, page 184) or citrulline to argininosuccinic acid (arrow 2, page 184), or the one needed for the conversion of argininosuccinic acid to arginine (arrow 3, page 185) may be missing.   Aminoaciduria is usually present.   The accumulation of ammonia in the tissues (brain) and blood is toxic.

PHENYLKETONURIA.   Phenylalanine hydroxylase is the missing enzyme (see arrow 1, page 179) in this disease.   Phenylalanine therefore cannot be converted to tyrosine.   Since the major pathway of phenylalanine degradation is blocked, this amino acid and its derivatives accumulate in the blood and are excreted in the urine.

In the absence of phenylalanine hydroxylase, phenylalanine cannot serve as a source for tyrosine, and so there is less tyrosine available for the formation of melanin and catecholamines.   The increased phenylalanine concentration in blood and the nervous tissue may produce brain damage.

ALBINISM.   This is a hereditary disease characterized by pale skin and hair and a general absence of pigment formation, often connected with poor vision.   The enzyme tyrosinase is deficient (see arrow 1, page 180) and therefore melanin formation is poor or absent.

ALKAPTONURIA.   The catabolism of phenylalanine and tyrosine is interrupted at the homogentisic acid level because of the absence of the enzyme homogentisic acid oxidase (see arrow 2, page 179).   The presence of homogentisic acid in the urine is characteristic of this disease; urine containing homogentisic acid darkens if it is left standing for several hours.

PROTEIN TURNOVER IN THE WHOLE MAMMALIAN ORGANISM

The introduction of radioactive isotopes as tracers for metabolic experiments has proved to be the most important tool in exploring anabolic and catabolic processes in tissues and in the whole organism. This technique shows conclusively that under normal conditions the tissue proteins of living organisms (like other biological compounds that serve as building blocks of living organisms) are constantly being synthesized and degraded. It is the dynamic equilibrium between synthesis and degradation rather than the prolonged static conservation of the synthesized proteins that ensures the proper function of the individual tissues and of the whole animal.

Turnover is a measure of this dynamic state. It is the rate by which a given amount of tissue protein (or other biological material) is replaced by newly synthesized proteins. The fact that the newly synthesized protein is a replacement for the "old" protein involves two qualifications. One is that the "old" and newly synthesized proteins are identical and the second is that the "old" protein is degraded. Therefore the turnover is a measure of degradation and synthesis of a given material in the state of dynamic equilibrium.

Experiments have shown that the turnover of some proteins which are part of the structure of a cell (red cell hemoglobin) or of a subcellular particle (mitochondrial protein) is often but not necessarily connected to the life span of the cell or subcellular particle. On the other hand, the turnover of serum proteins and some free proteins of the cellular protoplasm is apparently a random process. The turnover of proteins incorporated into cellular or subcellular structures may involve the "death" or decomposition of the structure itself as well as the birth of new cells or subcellular structures. The random turnover of free proteins is connected with the destruction of protein molecules rather than the death of whole cells. Both types of turnover occur in mammalian organisms. The destruction of hemoglobin follows the death of the red cell and the synthesis of new hemoglobin is connected with the formation of new red cells. Serum α and β globulins as well as fibrinogen are synthesized in liver cells, but their turnover is not connected with any large-scale disintegration or new formation of liver cells.

The method of studying protein turnover is to introduce a labeled precursor (amino acid) into the tissue or organism and measure the appearance of radioactivity as a function of time in the tissue or protein which is the subject of the investigation. It is assumed that the labeled amino acid does not alter the normal metabolism. It is also assumed that the labeled amino acid enters a common pool with similar unlabeled amino acids present in the cell and that unlabeled and labeled forms of a given amino acid are utilized randomly with equal probability for protein synthesis.

This general approach has provided us with an insight into how fast the proteins of individual tissues are degraded and replaced or, in other words, the life span of the tissue proteins. The turnover rate of proteins in various tissues shows great variation. Collagen is the most inert protein in adult mammals. Extremely slow (or nonexisting) protein turnover was found with collagen obtained from tendons and from large blood vessels. The turnover of muscle proteins is relatively slow. Proteins in the sarcoplasm appear to be replaced randomly. The half-life of muscle aldolase and muscle phosphorylase is estimated to be about 50 days. According to some tests the proteins of the myofibrils have a half-life of 30 to 80 days and perhaps their breakdown is connected with the life span of the myofibrils themselves. Since the muscle tissue constitutes a considerable portion of the body weight quantitatively, the protein turnover in this tissue at any time is a considerable fraction of the total protein turnover of the whole animal.

Proteins in the nervous system show great differences in turnover rates. About 50 per cent of the proteins in the central nervous system appear to have a half-life of approximately 150 days; another considerable portion of the proteins have a half-life of 16 days. Amino acid incorporation was measured in different sections of the nervous system. The incorporation was fastest into areas of high cellular density, such as cerebellum and cortex, and slowest in peripheral nerves.

The proteins of the absorbing epithelium as well as the cells themselves have very fast turnover rates. Their half-lives vary between 8 and 15 days at various portions of the digestive tract. The turnover rate of the intestinal smooth muscle and collagen is not well known but it is almost certain that the fast turnover rate of intestinal proteins can be attributed to the replacement of the epithelial cells.

The protein turnover rate in various organs such as liver, pancreas, and lungs varies widely from protein to protein. Some proteins in these organs have a half-life of 2 to 10 days, whereas some others (nucleohistones) appear to "live" for years.

Plasma Proteins

The proteins of the plasma are synthesized outside the vascular bed. Albumin, α and β globulins, and fibrinogen are synthesized in the liver and γ globulin is synthesized in the reticuloendothelial system. The synthesis of albumin, α and β globulins, and fibrinogen is apparently a random process not connected with replacement of liver cells. On the other hand, there are some indications that the synthesis of γ globulin is connected with a change in the cell population of the reticuloendothelial system.

Synthesis of serum albumin was studied in vitro with liver slices. It had been established that radioactive albumin appeared in the incubation media after a lag period. Usually 2 to 3 minutes were required after addition to the labeled amino acid. Since the labeled albumin was shown to be present in the liver cell instantaneously after the application of labeled amino acid, the lag period was interpreted as the time needed for release of albumin from an intracellular site. In vivo experiments show about 20 minutes' lag between the injection of tagged amino acid and the appearance of labeled albumin. The liver slices in vitro are capable of synthesizing serum albumin at a rate comparable to the albumin synthesis observed in the whole organism. Protein synthesis in the liver is probably controlled by the serum protein concentration since liver slices perfused with fibrinogen-free blood produced six times as much fibrinogen as those perfused with blood containing fibrinogen in normal concentrations.

It has been mentioned previously that while most serum proteins are synthesized in the liver, γ globulins are produced in the reticuloendothelial system. The stimulus for the synthesis of γ globulin is provided by a compound which is generally, but not always, exclusively a protein. This compound, defined functionally as an antigen, is foreign for the organism.

γ-Globulin concentration in the serum of a newborn is about 0.3 gm. per 100 ml. and it usually takes about 2 years to reach 0.8 gm. per 100 ml., the normal adult value. If an antigen is introduced into the organism, antibody (γ globulin) production usually starts after a 3 to 5 day latent period. After the latent period the antibody concentration rises exponentially during the next 4 to 5 days, may remain constantly high for several weeks or months, and finally diminishes.

The intracellular synthesis of other serum proteins takes only a few minutes and the secretion of proteins by a cell is usually completed within an hour. For these reasons the long latent period between the injection of antigens and the appearance of antibodies must have a cause other than the actual synthesis and secretion of the antibody molecule. It is supposed that the latent period is needed for the formation and multiplication of "immunologically competent cells" which are capable of antibody production. The central problem of antibody synthesis is the elucidation of the mechanism responsible for the recognition of the three-dimensional structure of the antigen in the antibody-synthesizing cells (Chapter 8).

REFERENCES

1. Broquist, H. P., and Trupin, J. S.: Amino acid metabolism. Ann. Rev. Biochem. *35*: 231–274 (1966).
2. Harris, H.: Human Biochemical Genetics. New York, Cambridge University Press, 1966.
3. Meister, A.: Biochemistry of the Amino Acid. 2nd edition. Vols. 1 and 2. New York, Academic Press, 1965.
4. Munro, H. N., and Allison, J. B. (eds.): Mammalian Protein Metabolism. Vols. 1 and 2. New York, Academic Press, 1964.

10

METABOLISM OF NUCLEIC ACIDS

DIGESTION AND ABSORPTION OF NUCLEIC ACIDS

Foodstuffs containing nucleoproteins are attacked by the acid milieu in the stomach and the nucleoproteins are dissociated into their nucleic acid and protein components. The nucleic acids are not further digested in the stomach.

The enzymes necessary for the hydrolysis of nucleic acids are secreted by the pancreas and probably by the intestinal mucosa. The pancreas produces two powerful endonucleases. Ribonuclease hydrolyzes ribonucleic acids to oligoribonucleotides and deoxyribonuclease cleaves deoxyribonucleic acids to oligodeoxyribonucleotides. The further degradation of the oligonucleotides is facilitated by phosphodiesterases which act as exonucleases and liberate mononucleotides. The mononucleotides are further cleaved to nucleoside and phosphate by the action of phosphatases. Both phosphodiesterase and phosphatase are believed to be elaborated by the intestinal mucosa. Nucleoside and phosphate appear to be the end products of the intestinal degradation of both ribonucleic acids and deoxyribonucleic acids. Nucleosides are absorbed through the intestinal cells into the blood. Enzymes capable of the hydrolysis of nucleosides to bases and pentose have been found in various tissues of the mammalian organism. The specificity and precise mechanism of action of the various nucleosidases is not known.

BIOSYNTHESIS OF PURINES

A large body of experimental evidence proves conclusively that the mammalian organism is not dependent upon dietary purines and pyrimidines for the synthesis of its nucleic acids. As a matter of fact, small-molecular-weight precursors (e.g., formic acid, glycine, glutamine) are used preferentially for the synthesis of purines and pyrimidines.

Although in the subsequent discussion the names of the participating enzymes will not be mentioned, it must be emphasized that the various steps of purine and pyrimidine biosynthesis are catalyzed by specific enzymes. Figure 10-1 shows the most important precursors of the various carbon and nitrogen atoms of the purine ring.

It may be seen (Fig. 10-1) that glycine serves as a precursor of C_4 and C_5 as well as N_7. The C atoms of formate are incorporated into C_2 and C_8, while carbon dioxide supplies C_6 for the purine ring. N_3 and N_9 are derived from the amide groups of glutamine, and the amino group of aspartic acid and glutamine donates N_1. Analysis of the intermediates of purine biosynthesis shows that mononucleotides are formed first. The mononucleotides then

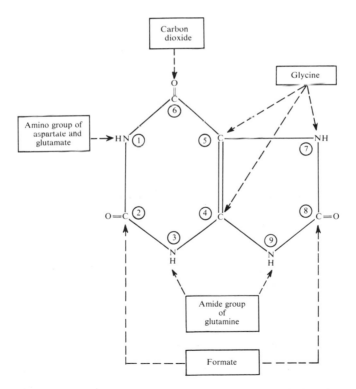

Figure 10-1. Small-molecular-weight precursors for purine ring synthesis in the mammalian organism.

either participate in the nucleic acid synthesis after further phosphorylation or are dephosphorylated to nucleosides and finally degraded to bases and pentose.

1. The first step in purine biosynthesis is the formation of 5-phosphoribosyl pyrophosphate (PRPP). This involves the transfer of the terminal pyrophosphate group of an ATP molecule to a ribose phosphate molecule.

$$
\begin{array}{c}
\text{OPO}_3\text{H} \\
|\\
\text{CH}_2 \\
|\quad\diagdown\text{O}\diagdown \\
\text{HC}\quad\text{HCOH} + \text{ATP} \\
\diagdown\diagup \\
\text{HC}-\text{CH} \\
\text{HO}\quad\text{OH}
\end{array}
\xrightleftharpoons{enzyme}
\begin{array}{c}
\text{OPO}_3\text{H} \\
|\\
\text{CH}_2 \qquad\quad\text{OH}\qquad\text{OH} \\
|\quad\diagdown\text{O}\diagdown\qquad|\qquad| \\
\text{HC}\quad\text{HC}-\text{O}-\text{P}-\text{O}-\text{P}-\text{OH} + \text{AMP} \\
\diagdown\diagup\qquad\quad\|\qquad\| \\
\text{HC}-\text{CH}\qquad\text{O}\qquad\text{O} \\
\text{HO}\quad\text{OH}
\end{array}
$$

Ribose 5'-phosphate   Phosphoribosyl pyrophosphate (PRPP)

2. The second step involves the transfer of an amino group from glutamine to PRPP.

$$
\text{Glutamine} + \text{PRPP} \xrightleftharpoons{enzyme}
\begin{array}{c}
\text{OPO}_3\text{H} \\
|\\
\text{CH}_2 \\
|\quad\diagdown\text{O}\diagdown\quad\diagup\text{NH}_2 \\
\text{HC}\qquad\text{CH} \\
\diagdown\diagup \\
\text{HC}-\text{CH} \\
\text{HO}\quad\text{OH}
\end{array}
+
$$

5'-Phosphoribosylamine

Pyrophosphate + Glutamate

3. 5-Phosphoribosylamine is a very unstable product which in the presence of ATP, Mg^{++}, and the proper enzyme conjugates glycine.

5-Phosphoribosylamine + glycine + ATP $\xrightleftharpoons{enzyme}$

$$
\begin{array}{c}
\text{OPO}_3\text{H} \\
|\\
\text{CH}_2 \qquad\quad\text{O} \\
|\quad\diagdown\text{O}\diagdown\quad\text{H}\quad\|\quad\text{H} \\
\qquad\qquad\text{N}-\text{C}-\text{C}-\text{NH}_2 + \text{ADP} + \text{Phosphate} \\
\text{HC}\qquad\text{CH}\qquad\qquad\text{H} \\
\diagdown\diagup \\
\text{HC}-\text{CH} \\
\text{HO}\quad\text{OH}
\end{array}
$$

Glycinamide ribonucleotide

4. The glycinamide ribonucleotide in the presence of a proper enzyme and a formyl donor is formylated to become N-formylglycinamide ribonucleotide.

Glycinamide ribonucleotide + N-5, N-10-Anhydroformyl-tetrahydrofolic acid + $H_2O \xrightarrow{\text{enzyme}}$

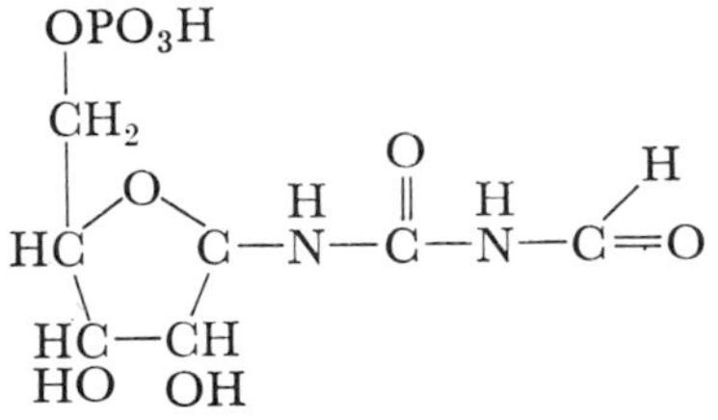

$$\begin{array}{c}
\text{OPO}_3\text{H} \\
| \\
\text{CH}_2 \\
\end{array}$$

HC̈   C—N—C—N—C=O + Tetrahydrofolic acid + H^+

HC—CH
HO OH

N-Formylglycinamide ribonucleotide

5. The next step involves an ATP-dependent amination of the N-formylglycinamide ribonucleotide using glutamine as the source of the amino group. The enzyme catalyzing this reaction is readily inhibited by azaserine and 6-diazo-5-oxonorleucine, structural analogs of glutamine.

Formylglycinamide ribonucleotide + Glutamine + ATP + $H_2O \xrightarrow{\text{enzyme}}$

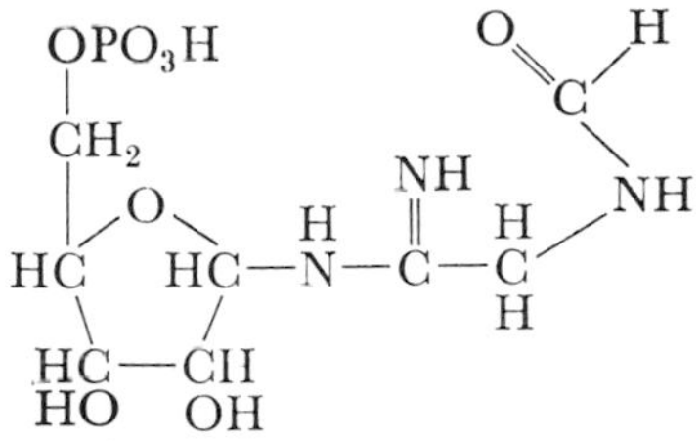

$$\begin{array}{c}
\text{OPO}_3\text{H} \\
| \\
\text{CH}_2 \\
\end{array}$$

HC HC—N—C—C   + Glutamic acid + ADP + Phosphate

HC—CH
HO OH

Formylglycinamidine ribonucleotide

6. ATP is again involved in the subsequent step of purine biosynthesis; it is the energy donor for a dehydration reaction leading to the closure of the imidazole ring.

Formylglycinamide ribonucleotide + ATP $\xrightarrow{\text{enzyme}}$

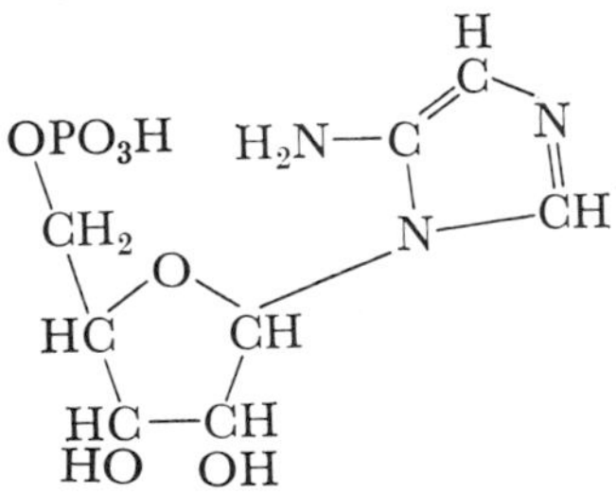

$$\begin{array}{c}
\text{OPO}_3\text{H} \\
| \\
\text{CH}_2 \\
\end{array}$$

HC   CH   + ADP + Phosphate

HC—CH
HO OH

5′-Aminoimidazole ribonucleotide

7. Carbon dioxide or dissolved carbonate is used for the carboxylation of 5-aminoimidazole ribonucleotide.

5-Aminoimidazole ribonucleotide $+ CO_2 \xrightleftharpoons{\text{enzyme}}$

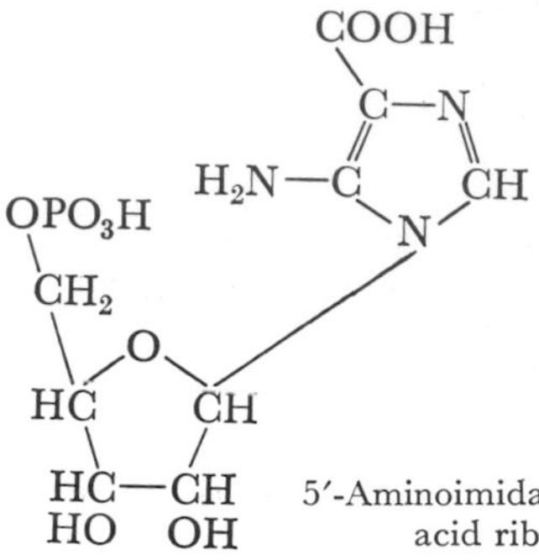

5′-Aminoimidazole-4-carboxylic acid ribonucleotide

8. The next step in the synthesis of purines is an ATP-dependent amide linkage formation between 5-aminoimidazole-4-carboxylic acid ribonucleotide and aspartic acid.

5-Aminoimidazole-4-carboxylic acid $+$ aspartic acid $+$ ATP $\xrightleftharpoons{\text{enzyme}}$

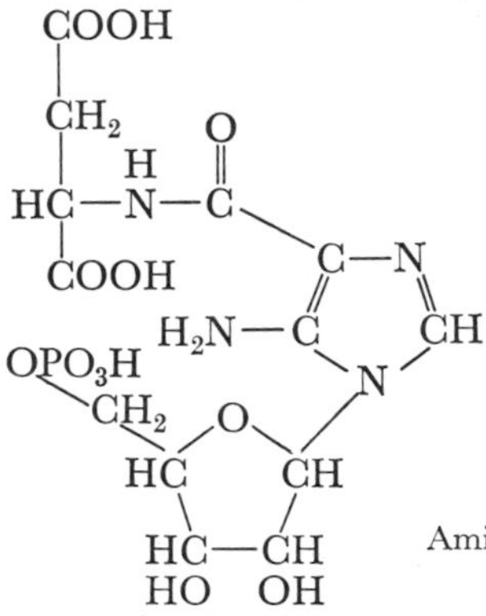

$+$ ADP $+$ Phosphate

Aminoimidazole-4-N-succinocarboxamide ribonucleotide

9. Fumaric acid is eliminated from the 5-aminoimidazole-4-N-succino-carboxamide ribonucleotide in the following step:

5-Aminoimidazole-4-N-succinocarboxamide ribonucleotide $\xrightleftharpoons{\text{enzyme}}$

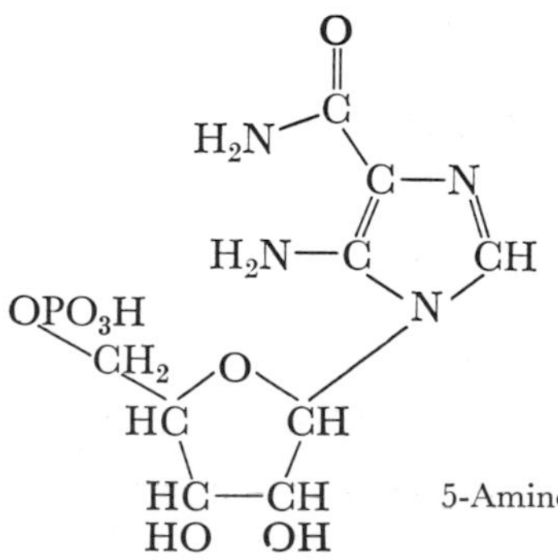

$+$ Fumarate

5-Aminoimidazole-4-carboxamide ribonucleotide

10. The formylation of 5-aminoimidazole-4-carboxamide ribonucleotide by N-10-formyltetrahydrofolic acid is the next synthetic step in the biosynthesis of inosinic acid.

5-Aminoimidazole-4-carboxamide tribonucleotide+N-10-formyltetrahydro-folic acid $\xrightarrow[]{\text{enzyme}}$

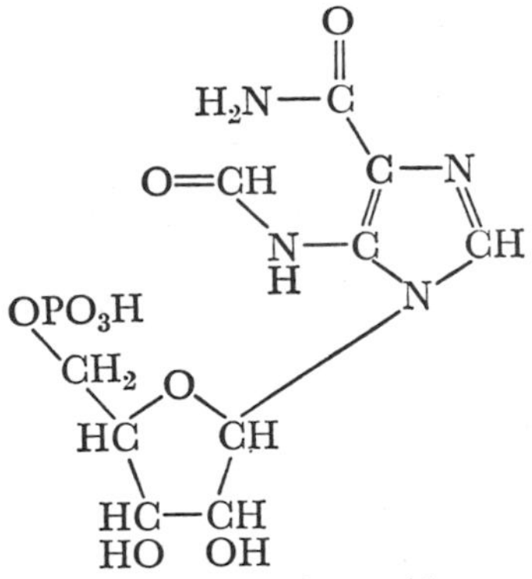

5-Aminoimidazole-4-carboxamide

$+$ Tetrahydrofolic acid

11. The last step is an enzyme-catalyzed cyclization of the formamido compounds to inosinic acid.

5-Formaminoimidazole-4-carboxamide ribonucleotide $\xrightarrow[]{\text{enzyme}}$

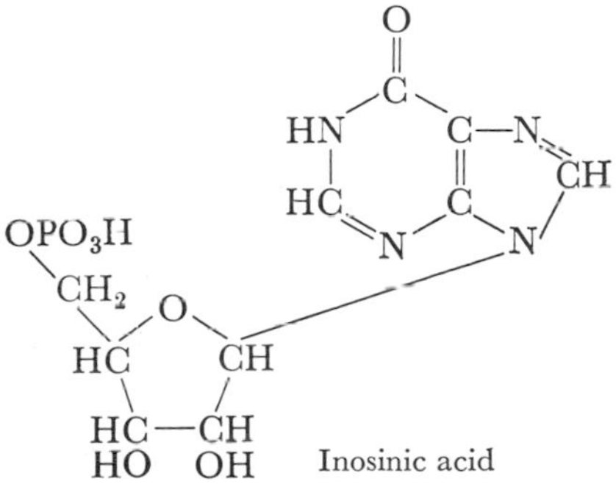

Inosinic acid

Inosinic acid is a common intermediate in the synthesis of adenylic acid and guanylic acid.

Adenylic acid synthesis takes place in two steps. The first involves a GTP-dependent amino group transfer from aspartic acid to inosinic acid.

Inosinic acid + aspartic acid + GTP $\xrightarrow[]{\text{enzyme}}$

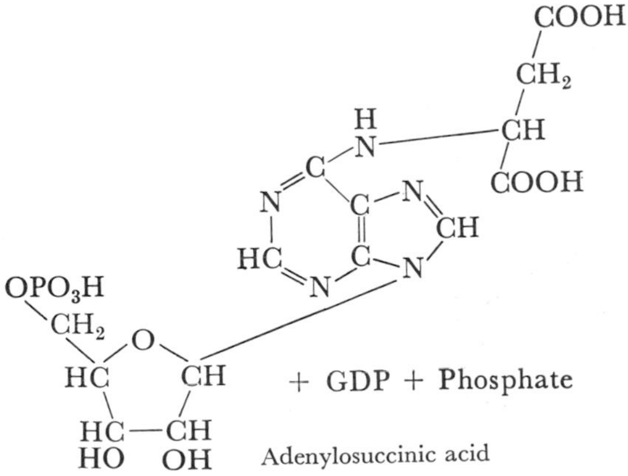

$+$ GDP + Phosphate

Adenylosuccinic acid

The second step consists of an elimination of a fumaric acid molecule from the adenylosuccinic acid to yield adenylic acid.

Adenylosuccinic acid $\xrightarrow{\text{enzyme}}$

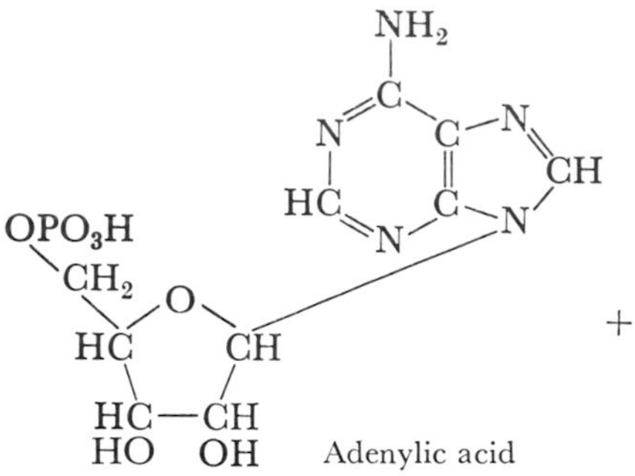

+ Fumaric acid

Adenylic acid

The formation of guanylic acid from inosinic acid also proceeds in two steps. In the first DPN oxidizes inosinic acid to xanthylic acid.

Inosinic acid + DPN + H_2O $\xrightarrow{\text{enzyme}}$

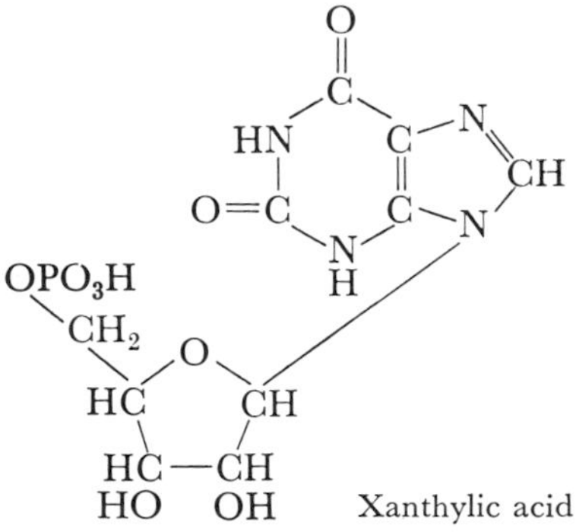

+ DPNH + H^+

Xanthylic acid

Xanthylic acid is then aminated to guanylic acid in the presence of ATP, with glutamine used as the primary nitrogen donor.

Xanthylic acid + Glutamine + ATP $\xrightarrow{\text{enzyme}}$

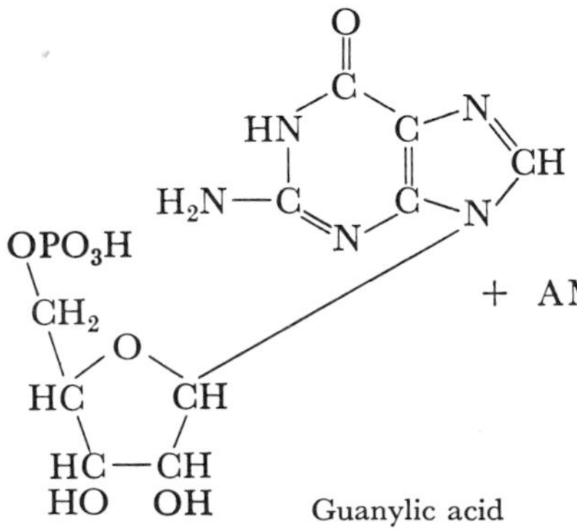

+ AMP + Glutamic acid + Pyrophosphate

Guanylic acid

BIOSYNTHESIS OF PYRIMIDINES

In the biosynthesis of the pyrimidines, as in the case of the purines, small-molecular-weight precursors play an important role. A major difference between the purine and pyrimidine biosynthesis is the point at which glycosidic bond formation takes place between the ribose phosphate and the various precursors participating in the synthesis of the purine or pyrimidine ring. In the case of the pyrimidines the glycoside bond is formed only after the pyrimidine ring is closed.

Figure 10-2 shows the small-molecular-weight precursors of orotic acid. It may be seen that N_1 is derived from NH_3; C_2 from CO_2; and N_3, C_4, C_5, and C_6 from aspartic acid.

1. Carbamyl phosphate plays an important role in pyrimidine biosynthesis. Therefore the formation of carbamyl phosphate, using NH_3, ATP, acetyl glutamate, and Mg^{++}, may be mentioned as the first step in the biosynthesis of the pyrimidine ring.

$$CO_2 + NH_3 + 2\ ATP \xrightarrow[\text{acetylglutamate}]{\text{enzyme}} H_2N\overset{\displaystyle O}{\overset{\displaystyle \|}{-C}}-OPO_3H + 2\ ADP +$$

Phosphate

Carbamyl phosphate

2. The carbamyl group of carbamyl phosphate is transferred next to aspartic acid.

Carbamyl phosphate + aspartic acid $\xrightleftharpoons{\text{enzyme}}$

$$
\begin{array}{ccc}
& COOH & \\
H_2N & \backslash\ CH_2 & \\
| & | & + \text{Phosphate} \\
O{=}C & C{-}H & \\
\backslash & \diagup\ \backslash & \\
N & COOH & \\
H & &
\end{array}
$$

Carbamyl aspartate

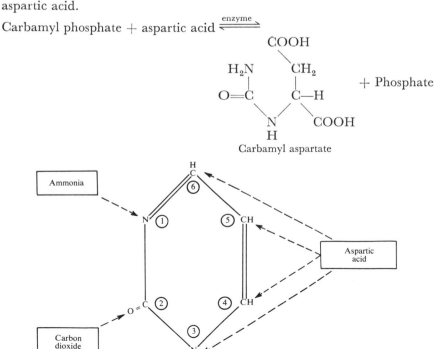

Figure 10-2. Small-molecular-weight precursors for pyrimidine ring synthesis in mammals.

3. Carbamyl aspartate undergoes an enzymatic dehydration which closes the ring.

$$\text{Carbamyl aspartate} \xrightleftharpoons{\text{enzyme}}$$

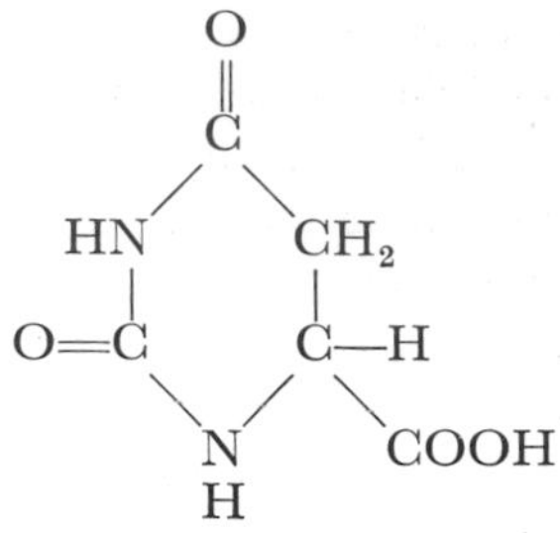

Dihydroorotic acid

$$+ \; H_2O$$

4. The synthesis of the pyrimidine ring is completed by an NAD-linked oxidation producing orotic acid.

$$\text{Dihydroorotic acid} + \text{NAD} \xrightleftharpoons{\text{enzyme}}$$

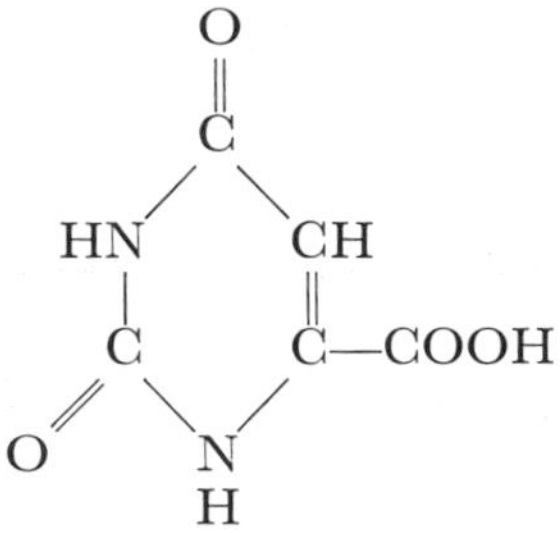

Orotic acid

$$+ \; NADH + H^+$$

5. In the next step the glycosidic bond is formed between ribose-5-phosphate and orotic acid. It is phosphoribosyl pyrophosphate rather than ribose-5-phosphate which actually participates in the condensation reaction with orotic acid. The result is orotidine-5′-phosphate, a pyrimidine nucleotide.

$$\text{PRPP} + \text{Orotic acid} \xrightleftharpoons{\text{enzyme}}$$

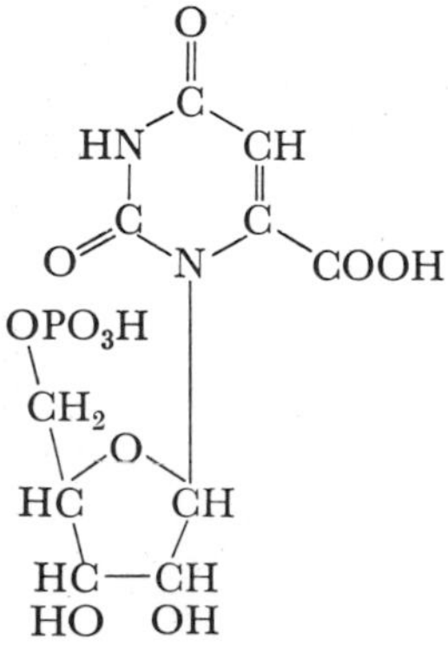

$$+ \; \text{Pyrophosphate}$$

Orotidine-5′-phosphate

6. Uridylic acid is formed from orotidine-5′-phosphate by decarboxylation.

$$\text{Orotidine-5'-phosphate} \underset{\text{enzyme}}{\rightleftharpoons}$$

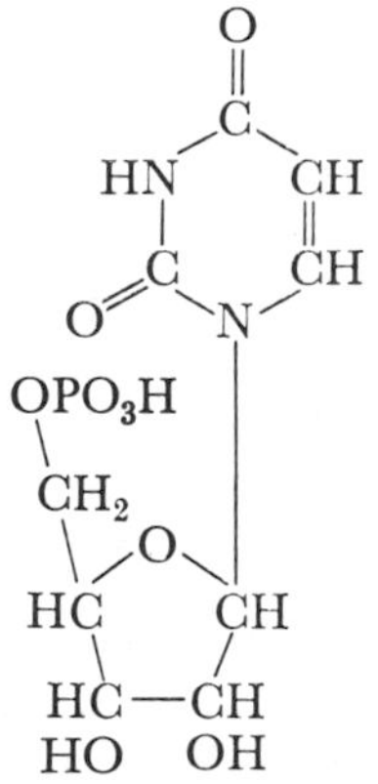

Uridylic acid

$+ \ CO_2$

The formation of cytidine nucleotide from uridylic acid involves two successive phosphorylation steps leading to the formation of UDP and UTP, respectively.

7. $\qquad$ UMP* + ATP $\underset{\text{enzyme}}{\rightleftharpoons}$ UDP + ADP

8. $\qquad$ UDP + ATP $\underset{\text{enzyme}}{\rightleftharpoons}$ UTP + ADP

9. Ammonia is the N donor for the amination of UTP to yield CTP.

$$\text{UTP} + \text{NH}_3 + \text{ATP} \underset{\text{enzyme}}{\rightleftharpoons} \text{CTP} + \text{ADP}$$

Uridylic acid occurs only in RNA, cytidylic acid is a component of both RNA and DNA, and thymidylic acid is present only in DNA. It has been shown that uniformly labeled cytidine and uridine are incorporated into the deoxycytidylic and thymidylic acid without cleavage of the glycosidic bond. Therefore, the reduction of ribose to deoxyribose may occur either on the nucleotide or nucleoside level. There is good indication that nucleotide diphosphates are used primarily for the reduction of the sugar moiety and the enzymes require NADPH, ATP, Mg^{++}, and a sulfhydryl protein cofactor, called thioredoxin.

Orotic acid, uracil, cytidine, uridine, and cytosine may serve as precursors for thymidylic acid. There is some evidence to indicate that the reduction of ribose for thymidylic acid synthesis proceeds best on CDP-UDP level. The dUDP is dephosphorylated to dUMP. The methyl group for the thymidylate synthesis may come from formate, serine, or glycine. N-5, N-10-Methylene tetrahydrofolic acid and Mg^{++} and the proper enzyme are needed for the methylation reaction.

---

* XMP, XDP, XTP = nucleoside monophosphate, diphosphate, and triphosphate.

Deoxyuridylic acid + N-5, N-10-Methylene tetrahydrofolic acid $\xrightleftharpoons{\text{enzyme}}$

$$
\begin{array}{c}
\text{H} \\
\text{N}
\end{array}
$$

OPO$_3$H   O=C   C=O

CH$_2$ N   C—CH$_3$

O

HC   CH   C

HC——CH

HO   OH

+ Dihydrofolic   acid

Thymidylic acid

These metabolic pathways constitute the main route for purine and pyrimidine biosynthesis. Alternative synthetic mechanisms are also known, but since their biological significance is not clear they are not discussed in this text.

SYNTHESIS OF MACROMOLECULAR NUCLEIC ACIDS

In the first two sections of this chapter we have shown the most important pathways for the synthesis of purine and pyrimidine mononucleotides from small-molecular-weight precursors. These mononucleotides may be subjected to enzymatic transphosphorylations and give rise to nucleotide diphosphates and triphosphates.

Nucleotide monophosphate + ATP ⇌ Nucleoside diphosphate + ADP

Nucleotide diphosphate + ATP ⇌ Nucleoside triphosphate + ADP

These transphosphorylations are important for the synthesis of macromolecular nucleic acids since most polymerizing enzymes require nucleoside triphosphates for the polymerization reaction.

Synthesis of Macromolecular DNA

The enzyme that catalyzes the synthesis of macromolecular DNA is called DNA polymerase. It uses deoxynucleoside triphosphates as substrates; deoxynucleoside diphosphates and deoxynucleoside monophosphates are not polymerized. The reaction is reversible and requires the presence of divalent cations (Mg^{++}, Mn^{++}). All four nucleoside triphosphates have to be present; the absence of any one of them abolishes the reaction. A DNA primer is also necessary for the polymerization to occur; in its absence the polymerization is depressed 1000 to 100,000 fold. RNA cannot serve as primer. The overall scheme of the DNA polymerization is the following:

$$
a[dATP] + b[dGTP] + e[dCTP] + c[dTTP] \xrightleftharpoons[\text{+Mg}^{++}\text{, DNA primer}]{\text{enzyme}}
$$

$$
\begin{bmatrix} a[dATP] \\ b[dGTP] \\ e[dCTP] \\ c[dTTP] \end{bmatrix}_{\text{DNA}} \quad (a + b + e + c) \text{ Pyrophosphate}
$$

Addition of pyrophosphate in large concentration promotes depolymerization. The polymerization proceeds by formation of a phosphate ester link between the free 3'-hydroxyl group of the terminal ribose in the growing chain and the 5'-phosphoric acid group of the nucleotide which is being attached to the chains.

Although the complete base sequence analysis is still a somewhat difficult problem there is indirect evidence that the primer DNA and the synthesized DNA may have similar base sequences. The so-called "nearest-neighbor" analysis provides information regarding the absolute frequency of the individual dinucleotide combinations in a DNA molecule. The 16 possible dinucleotide combinations are:

$$A-A,* \quad G-G, \quad C-C, \quad T-T$$
$$A-G, \quad G-A, \quad C-A, \quad T-A$$
$$A-C, \quad G-C, \quad C-G, \quad T-G$$
$$A-T, \quad G-T, \quad C-T, \quad T-C$$

The purpose of the nearest-neighbor analysis is to disclose the absolute frequency of any given dinucleotide combination in a DNA molecule. If the absolute frequencies of various dinucleotide combinations in two DNA molecules are exactly the same, this is accepted as good evidence that the nucleotide sequence of the two DNA molecules is identical. Figure 10-3*B* shows a tetranucleotide portion of a hypothetical DNA molecule synthesized by the DNA polymerase. In this case only the thymidine triphosphate contained labeled phosphorus atoms in its phosphoric acid residues. The other three nucleotide triphosphates used for this synthesis were unlabeled (Fig. 10-3*A*).

In the newly synthesized DNA molecule the labeled phosphoric acid is attached 5' position to the thymidines and 3' position to the left-side nearest neighbor of each thymidylic acid. With the proper enzyme it is possible to cleave all bonds of the DNA molecule connecting the phosphoric acid residues with the 5' atom of the nucleosides (Fig. 10-3*C*). After such hydrolysis the DNA is broken down to its constituent 3'-nucleotides. It is easy to see that the labeled phosphoric acid which was introduced as 5'-phosphate of thymidylic acid after the enzymatic hydrolysis became translocated to the left-side nearest neighbor of the thymidylic acid constituents of the DNA molecule.

In the case of the tetranucleotide depicted in Figure 10-3*B*, the labeled phosphoric acid participated in the 5'-3'-phosphate ester link formation between nucleotides C and T. After all the 5'-phosphoric acids were cleaved from their nucleosides (*C* in Figure 10-3), the labeled phosphate remained attached to nucleoside C. Generally this method will indicate how many times the labeled nucleotide was the right-side nearest neighbor of each of the four nucleotides (including itself) in the newly synthesized nucleic acid. If the enzymatic synthesis is carried out four times, with a different labeled nucleoside triphosphate in each experiment, this analysis will determine the

---

* A-A = adenylic acid + adenylic acid; T-C = thymidylic acid + cytidylic acid, and so on.

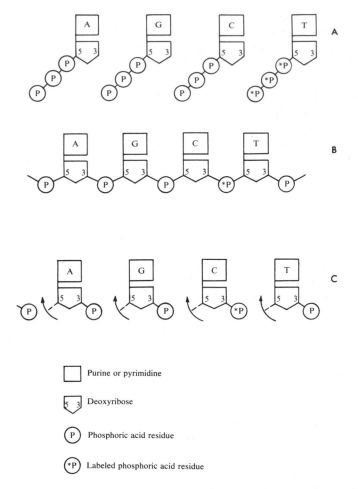

Figure 10-3. Schematic representation of nearest-neighbor analysis. *A*, Nucleoside triphosphate used for the DNA synthesis. *B*, A tetranucleotide segment of the newly synthesized DNA molecule. *C*, The 3′ nucleotide constituents of the tetranucleotide shown in *B*, after cleavage with the proper enzyme shown by the arrows. It may be seen that the labeled phosphorus was translocated from thymidine to its left-side nearest neighbor, cytidine.

absolute frequencies of all 16 dinucleotides in the nucleic acid molecule. This experimental procedure has produced strong evidence that the base sequence of the DNA molecule synthesized by DNA polymerase is similar to the base sequence of the primer DNA used for the synthetic process.

All these results would seem to suggest that the primer serves as a template and the synthesized DNA is a copy of the template. The proper base sequence is ensured by the acceptance or rejection of a given nucleotide by the template. Prior to the 3′-phosphoester bond formation the base of the template may form hydrogen bonds with the base of the proper nucleotide. Since only A-T and G-C pairing is possible, the proper nucleotide sequence may be to control by the template.

Physical properties of the newly synthesized DNA are also similar to but not absolutely identical with those of the primer. There is some evidence which indicates that in vitro the synthesized DNA may be a true copy of a certain portion of the primer but it need not be a perfect copy of the entire primer.

Synthesis of Macromolecular RNA

Several RNA-polymerizing enzymes have been discovered. The main differences between these enzymes lie in their respective substrate and primer requirements. A polynucleotide phosphorylase catalyzes the formation of RNA from the nucleoside diphosphates. The presence of every type of nucleoside diphosphates is not a requirement since this enzyme may catalyze the formation of RNA homopolymers (e.g., poly A) as well as various hetero-polymers (e.g., poly AGUC). Deoxynucleotide diphosphates are not poly-merized by this enzyme.

A primer RNA or DNA may dispense with the lag period that otherwise precedes the onset of the polymerization, but primer is not an absolute require-ment. The exact function of the primer in this case is not well understood but it is established that it does not serve as a template.

The RNA synthesized by polynucleotide phosphorylase resembles much more the original composition of the substrate mixture than the com-position of the template. The affinity of the substrate to the polynucleotide polymerase is very low ($K_m = 10^{-2} M/1$). The relatively low physiological concentration of free nucleoside diphosphate and the relatively high RNA and phosphate concentration in the cell would favor depolymerization rather than polymerization. Since this enzyme is not capable of "copying" a given template, it is unlikely that it is involved at all in the synthesis of "informational" RNA. On the other hand, this enzyme may play a role in the decomposition of informational RNA, probably to maintain a proper nucleotide pool within the cell.

Another RNA polymerase (transcriptase) appears to satisfy those requirements that would qualify it to be responsible for the synthesis of informational RNA. The overall reaction is indicated by the following equation:

$$a[ATP] + b[GTP] + c[CTP] + e[UTP] \xrightarrow[\text{Mg}^{++}, \text{ DNA primer}]{\text{enzyme}} \begin{bmatrix} a[AMP] \\ b[GMP] \\ c[CMP] \\ e[UMP] \end{bmatrix}_{RNA} + (a + b + c + e) \text{ Pyrophosphate}$$

This enzyme requires the presence of all four ribonucleoside triphosphates as well as a DNA primer. There is a large body of evidence which indicates that the DNA primer serves as a template and the RNA polymerase copies or, better, transcribes the DNA template into RNA. The base ratios of template and product, as well as the absolute frequencies of the various dinucleotides based on the nearest neighbor analysis, appear to be the same.

DNA template and newly synthesized RNA may form hybrid complexes which strongly indicate a good fit between these two molecules.

Besides these two enzymes, so-called RNA-dependent RNA polymerases or replicases have also been identified. These replicases may use RNA or DNA primers as templates and at least some of them may use deoxynucleoside polyphosphates as well as nucleotide polyphosphates as substrate. Such an enzyme may catalyze the formation of RNA-DNA hybrid complexes. The exploration of the biological role of these enzymes is a very interesting and challenging problem in molecular biology.

CATABOLISM OF DNA AND RNA

Enzymes capable of hydrolyzing DNA or RNA (nucleases) have been isolated from various tissues (e.g., pancreas, spleen, thymus). Some of these enzymes cleave only DNA (DNAase) while others split RNA molecules (RNAase). Endonucleases split randomly either the 3'- or the 5'-phosphoester linkages within the body of the nucleic acid molecule, leading to the formation of 5'- or 3'-oligonucleotides, respectively. Exonucleases preferentially split phosphoester linkages at one end of the molecule (e.g., ribonuclease type B splits 3'-phosphoester linkage starting at that end of the RNA molecule which carries the free 5'-phosphoric acid).

The combined action of endo- and exonucleases produces mononucleotides which are further degraded by the action of certain phosphatases (mononucleotidase) to nucleosides. The glycosidic bond of the nucleoside is further cleaved by nucleosidases to yield the free base and sugar. The majority of mononucleotidases and nucleosidases have no specific requirements regarding the base or sugar moiety of their respective substrates.

The catabolism of DNA and RNA will therefore yield bases, sugars, and phosphoric acid. The breakdown of the purine and pyrimidine bases proceeds as follows.

Adenine is deaminated primarily on the nucleoside level by an enzyme called adenosinase and further cleaved by nucleosidase. The products are hypoxanthine and sugar.

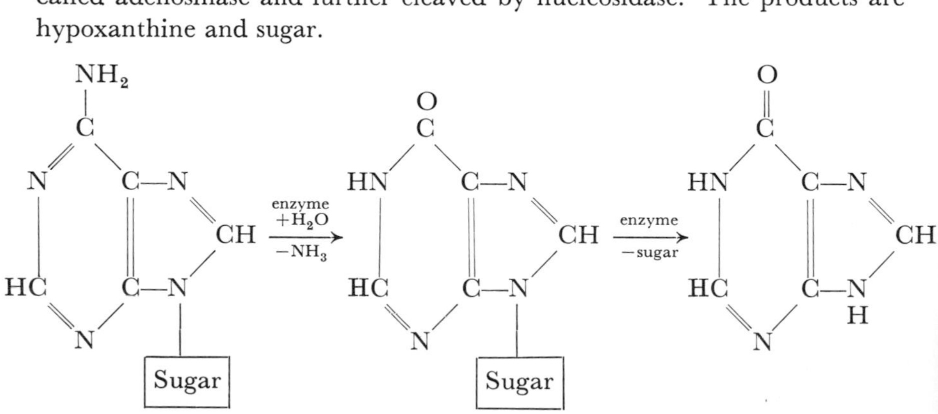

Adenosine   Hypoxanthine

Guanine is directly attacked by an enzyme called guanase and converted to xanthine.

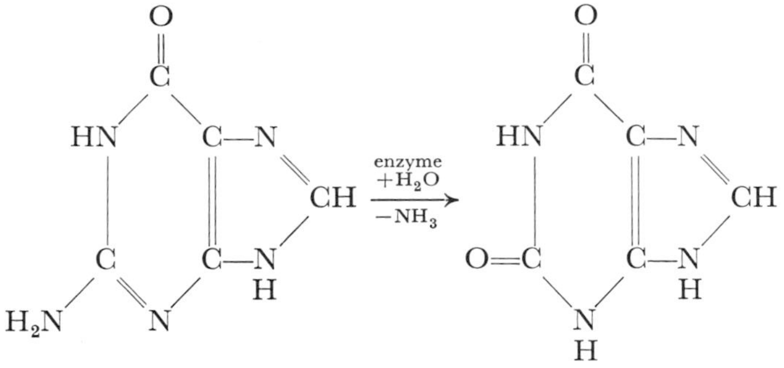

Guanine　　　　　　　　　　　Xanthine

Another enzyme called xanthine oxidase catalyzes the conversion of both hypoxanthine and xanthine to uric acid. Xanthine oxidase is a flavin enzyme which contains iron and molybdate and uses molecular oxygen for oxidation.

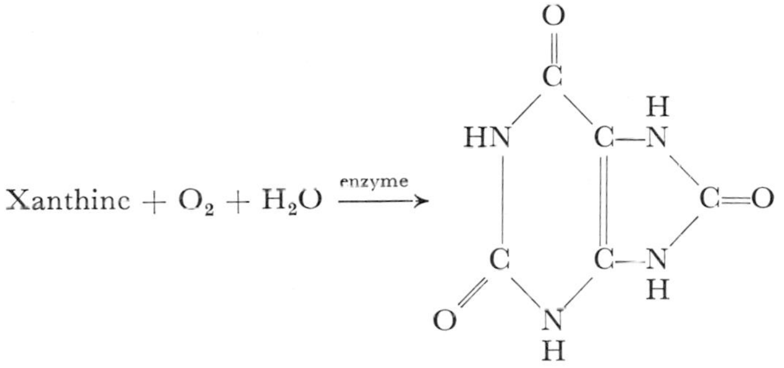

$$Xanthine + O_2 + H_2O \xrightarrow{enzyme}$$

Uric acid

Uric acid is excreted by the kidneys. Gout is a disease characterized by painful attacks in joints, formation of urate deposits in various tissues, and increased serum uric acid concentration. The cause of this disease is not fully understood.

The end products of uracil and thymine degradations are β-alanine and β-aminoisobutyric acids, respectively. The ureido carbon of the uracil molecule may appear in the respiratory carbon dioxide.

NUCLEIC ACID CONTENT AND TURNOVER IN VARIOUS TISSUES

DNA is in the cell nucleus and its absolute amount is quite constant in cells of similar ploidy. Cells which have small cytoplasm relative to their nuclei will have a relative larger DNA concentration, and those with large cytoplasm, smaller DNA concentration. Thymus, lymphoid tissue, bone marrow, and small intestine have large concentrations of DNA (100 μg. DNA phosphate per 100 mg. tissue). On the other hand, excitable tissues (muscle, nerve) or secreting tissues have relatively smaller DNA concentrations (50 μg. DNA phosphate per 100 mg.). The RNA concentration is high (50 μg. RNA phosphate per 100 mg. tissue) in those organs which are active in protein synthesis (reticuloendothelial system, liver, pancreas, small intestine) and low in muscle and nerve tissues.

Nucleic Acid Metabolism in Various Organs

Investigations that utilize the whole animal rather than tissue homogenates or partly purified enzymes to study the turnover of nucleic acids have contributed important results. It has been shown that among the purine bases adenine may serve as a precursor for RNA adenine and guanine and to a much smaller extent for DNA adenine and guanine. Guanine is also incorporated into the aforementioned nucleotides but to a much smaller extent than adenine. Adenosine, adenylic acid, guanosine, and guanylic acid are poorer precursors than their respective bases. Neither adenine nor guanine is utilized for the synthesis of pyrimidines. The pyrimidine bases uracil, cytosine, and thymine are very poorly or not at all utilized by various organisms as precursors for nucleic acids. Orotic acid is the only pyrimidine base that is well utilized in the synthesis of RNA and DNA pyrimidines in many mammalian organisms. On the other hand, cytidine, uridine, thymidine, and their corresponding nucleotides, if injected into the animal, are incorporated into the tissue nucleic acids.

Cytidine and cytidylic acid are best taken up into RNA and DNA cytosine but are also converted into the other pyrimidines. Cytosine and cytidylic acid are better precursors for the nucleic acid pyrimidines than uridine and uridylic acid. Deoxycytosine is incorporated only into DNA cytosine and thymine, and not into the RNA pyrimidines. Thymidine and thymidylic acid are taken up only into thymine. All these results clearly show that purines are not used for pyrimidine synthesis and pyrimidines are not suitable for the synthesis of purines.

Exogenous pentose is not utilized by the human and many other primate organisms for the synthesis of nucleic acids. On the other hand, exogenous phosphate is readily incorporated into RNA and DNA and has been extensively used for metabolic studies.

The incorporation of ^{32}P was found to be 15 to 60 times faster into the DNA of the intestines or bone marrow than into that of resting liver or muscle. Regenerating liver may show a 20 fold and hepatoma a 60 fold increase in the incorporation of ^{32}P into the DNA, as compared to resting liver. The increase of RNA synthesis under similar experimental conditions is relatively smaller than the increase of DNA synthesis.

It appears to be fairly well documented that three to five times more ^{32}P is taken up into the RNA than into the DNA of the resting rat liver. In regenerating rat liver and hepatoma the RNA ^{32}P/DNA ^{32}P ratio comes close to unity. Similar results were obtained with various tissues of several mammalian organisms. This would seem to indicate that the turnover of RNA in the resting organ is faster than that of the DNA. In proliferating tissues, however, the synthesis of the DNA is much faster than in resting tissues.

The rate of the incorporation of adenine and pyrimidine nucleosides is much smaller than the rate of incorporation obtained with phosphate or some other small-molecular-weight precursors (e.g., ^{14}C formation). This fact further substantiates the theory that under physiological conditions the mammalian organism uses small-molecular-weight precursors (formate, glycine, and so forth) for nucleic acid synthesis. No competition was found between adenine and formate ^{14}C with respect to their relative incorporation into RNA purine of rat intestine and liver if both precursors were employed simultaneously. There are some results indicating that if the so-called de novo nucleic acid synthesis (synthesis in from small-molecular-weight precursors) is inhibited, the nucleic acid synthesis from preformed purine or pyrimidine or both may increase several fold.

Among the vitamins folic acid plays the most important role in nucleic acid biosynthesis. Folic acid antagonists are effectively used to inhibit this process.

Inhibition of Nucleic Acid Synthesis

The most conspicuous biological difference between neoplastic and normal somatic cells is in their rate of division. Tumor cells divide more frequently than the normal cells of the same tissue. Since cell multiplication can not proceed without increased DNA, RNA, and protein synthesis, inhibition of these processes should affect the proliferating tissues much more than resting tissues. Similarly, bacterial cells multiply much faster than the somatic cells of the host organism. For these reasons it is possible that agents capable of interacting with DNA, RNA, or protein metabolism could be used as antitumor or antibacterial agents. In the last 20 years large numbers of purine and pyrimidine analogs have been synthesized.

6-Mercaptopurine   Thioguanine   8-Azaguanine

Among the purine analogs 6-mercaptopurine is probably most useful as an inhibitor of tissue growth. The growth-inhibitory effect of 6-mercapto-purine in some experimental tumors and bacteria is reversed by large concentrations of hypoxanthine, xanthine, adenine, and guanine. It has also been established that inosinic acid accumulates in the 6-mercaptopurine-inhibited tissues. Therefore apparently 6-mercaptopurine inhibits the conversion of inosinic acid to adenylic acid. The mechanism of inhibitory action is not quite as simple with thioguanine and 8-azaguanine. These compounds may inhibit the incorporation of guanine into the nucleic acid. At the same time they are incorporated into the nucleic acid of some tumors and bacteria and inhibit protein biosynthesis.

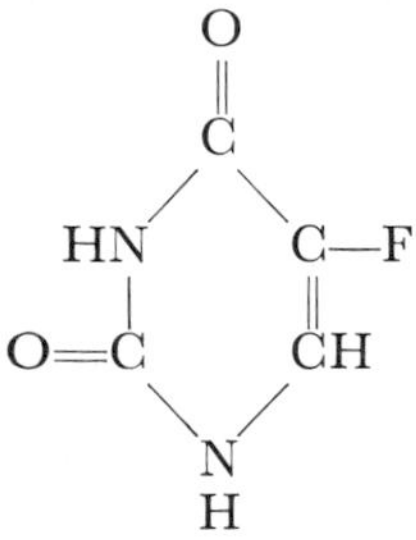

5-Fluorouracil 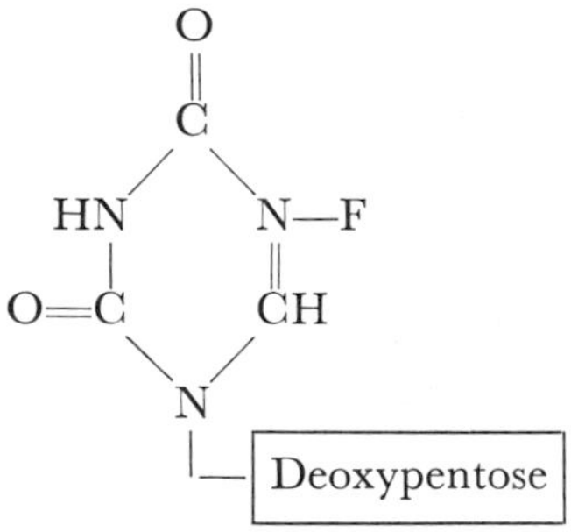   5-Fluorodeoxyuridine 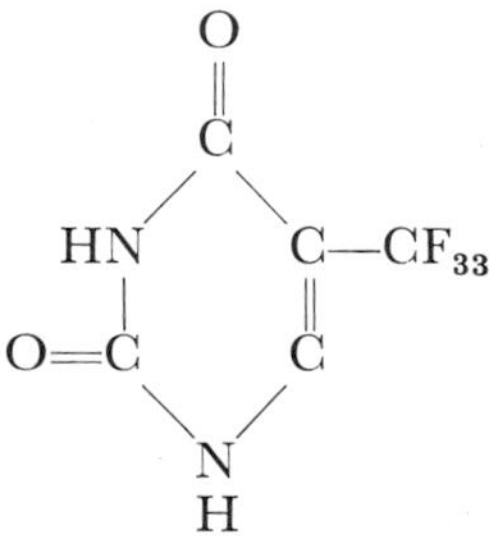   Trifluorothymidine

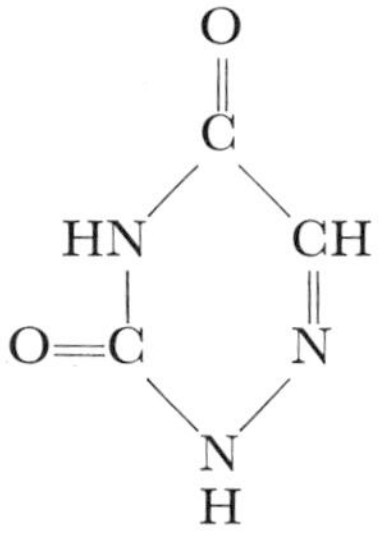

4(6)-Azauracil

5-Fluorouracil and its deoxynucleosides are probably the best anti-tumor agents among the antimetabolites of pyrimidine biosynthesis. Their antitumor action is probably due to the inhibition of thymidylic acid formation since both fluorodeoxyuridine and trifluorothymidine inhibit the action of thymidylate synthetase. Trifluorothymidylic acid is also incorporated into cellular DNA. It has been shown that DNA containing 11 per cent trifluorothymidylic acid is more sensitive to x-rays than DNA without this antimetabolite.

Azauracil inhibits the conversion of orotic acid to uridylic acid. Folic acid derivatives serve as coenzymes for one-carbon transfer reactions in two different steps of purine biosynthesis and in one step of pyrimidine biosynthesis. Therefore inhibition of the folic acid action may repress nucleic acid synthesis.

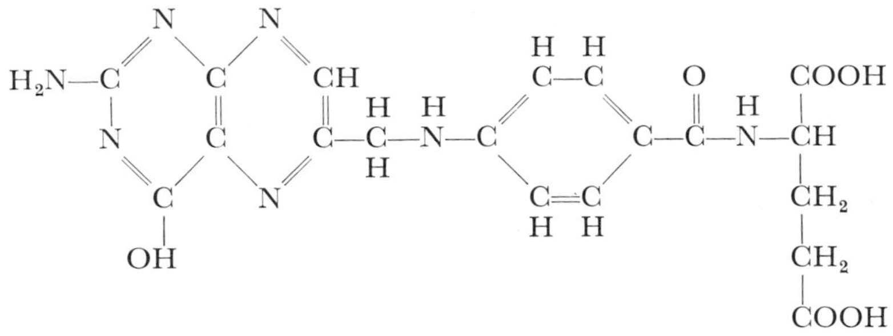

Folic acid (Pteroylglutamic acid)

Aminopterin inhibits reversibly an enzyme (folic acid reductase) that catalyzes the formation of the active coenzyme tetrahydrofolic acid. This latter compound is the active coenzyme in the one-carbon transfer. Aminopterin and some other folic acid antagonists produce temporary remissions in chronic leukemia of man and in a few experimental tumors of animals.

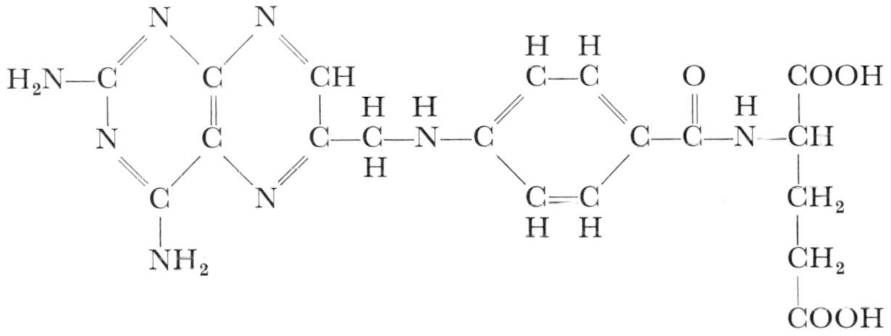

Aminopterin

Glutamine serves as nitrogen donor for the purine and pyrimidine rings. Some structural analogs of the glutamine may inhibit these aminations.

$$H_2N—CO—CH_2—CH_2—CHNH_2—COOH$$

Glutamine

Azaserine is able to repress the growth of some experimental tumors and bacteria. Its biological action is reversed by large concentrations of preformed purines or glutamine.

$$\overset{-}{N}=\overset{+}{N}—CH—CO—O—CH_2—CHNH_2—COOH$$

Azaserine

An antibiotic called meractinomycin has become an important tool in studying DNA-dependent RNA synthesis as well as the biosynthesis of proteins. The agent inhibits DNA-dependent RNA synthesis but does not repress DNA synthesis. It has been suggested that strong binding of meractinomycin to DNA interferes with the function of DNA as a template for RNA synthesis. Meractinomycin represses almost completely the RNA and protein synthesis in the cell.

REFERENCES

1. Blakley, R. L., and Vitols, E.: The control of nucleotide biosynthesis. Ann. Rev. Biochem. *37*: 201–224 (1968).
2. Chargaff, E.: Essays on Nucleic acids. New York, American Elsevier Publishing Co. 1963.
3. Chargaff, E., and Davidson, J. N. (eds.): The Nucleic Acids. 3 vols. New York, Academic Press, 1955 and 1960.
4. Lehman, I. R.: Deoxyribonucleases, Their Relationship to Deoxyribonucleic Acid Synthesis. Ann. Rev. Biochem. *36*: 645–668 (1967).
5. Progress in Nucleic Acid Research. New York, Academic Press.

INDEX

Page numbers in *italics* refer to illustrations.